CLAY CLARK'S

PODCAST DOMINATION

THE PROCESS AND PATH TO PODCAST SUCCESS

PODCAST DOMINATION

THE <u>PROVEN</u> <u>PATH</u> TO PODCAST SUCCESS

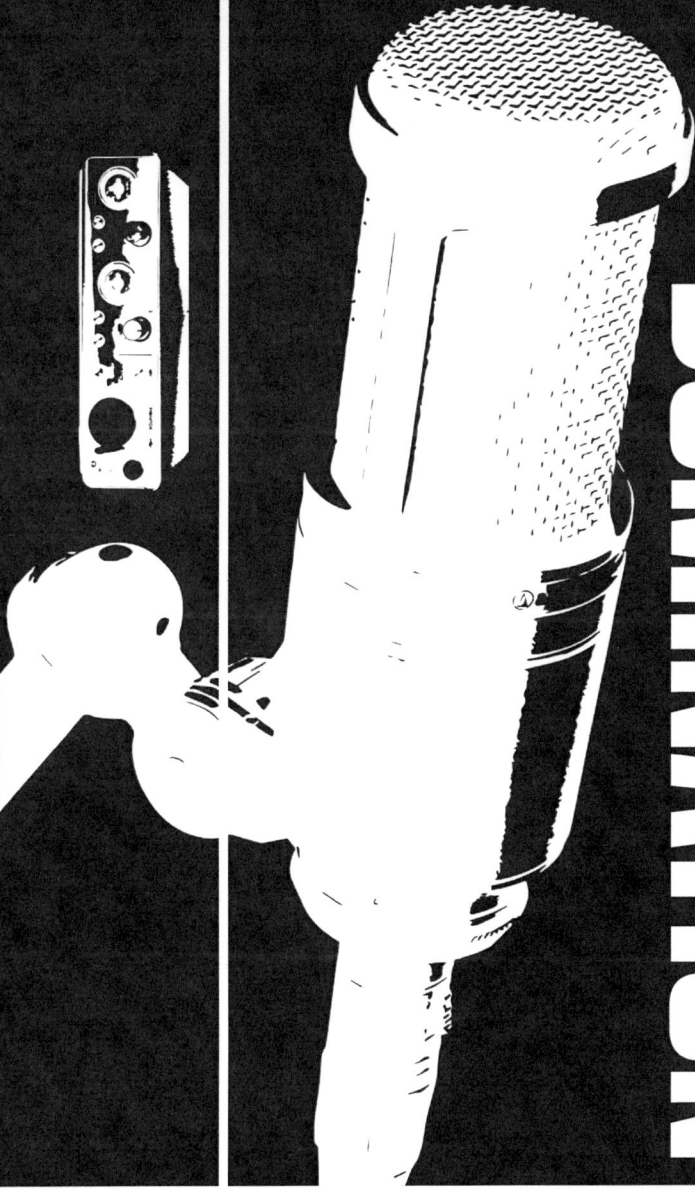

CLAY CLARK

"VISION WITHOUT EXECUTION IS HALLUCINATION."

THOMAS EDISON

(The creator of the modern light bulb, the inventor of recorded audio, the inventor of recorded video, and the founder of General Electric.)

THINGS YOU NEED TO KNOW AND FUN FACTS

Thus, far throughout my personal podcasting career, I have hit the top of the iTunes charts 6 times which is encouraging considering that as of 2018, the *Wall Street Journal* now reported that 26% (or 73 million) Americans now listen to podcasts at least once per month.

Throughout my career as a podcast/AM radio show host I have had the opportunity and pleasure to interview the following iconic industry leaders and more...(See the other guests I've interviewed today by visiting www.ThrivetimeShow.com):

The king of the culinary world - Wolfgang Puck

The lord of the leadership industry and iconic author of *The 21 Irrefutable Laws of Leadership* - John Maxwell

The sultan of the songwriting industry and a man who has written hit songs for the likes of Justin Bieber, Selena Gomez, Maroon 5, Linkin Park, Meghan Trainor, Cee Lo Green, Flo Rida, Jason Derulo, Charlie Puth, Demi Lovato, P!nk, Nicki Minaj and countless other top 40 music superstars - Ross Golan

The successful psychology superstar and *New York Times* best-selling author - Daniel Goleman

The fabulous personal finance guru and 9x *New York Times* best-selling author - David Bach

The 2x National Basketball Association Champion of the World, the number one overall draft pick, the 1995 NBA MVP, the 10-time NBA All-Star, the two-time Olympic Hall of Fame Inductee and super successful entrepreneur - David Robinson

The PR consultant of choice for Michael Jackson, Prince, Pizza Hut, Nike, etc. - Michael Levine

The co-founder of Ritz-Carlton - Horst Schultz

The former Executive Vice President of Walt Disney World Resorts who once was personally responsible for managing over 40,000 employees and 1,000,000 guests (customers) per week - Lee Cockerell

I would encourage you to **not** start a podcast until you are committed to not stopping. What? Yes. I repeat, I would encourage you to not launch a podcast until you are committed to not stopping. Diligence is the difference maker and you must commit to becoming so good that listeners, potential guests and sponsors simply cannot ignore you. Am I there yet? No. My goal is to faithfully produce the best practical podcast possible for the listeners of *The Thrivetime Show* 9 times per week while also committing myself to mastering my craft and to improving every week.

"BE SO GOOD THEY CAN'T IGNORE YOU."

STEVE MARTIN

(The award-winning actor, filmmaker, writer, comedian and musician behind the hit movies, *The Jerk, The Father of the Bride, The Father of the Bride II* which eventually awarded Steve Emmy, Grammy and American Comedy awards.)

☺

FUN FACTS

Moonrise - This show is hosted by Lillian Cunningham and is about the nuclear arms race during the time of the Cold War. Sounds like the kind of show that teaches those practical steps that you can go out and implement immediately TODAY...

The Joe Rogan Experience - This podcast is hosted by the stand up comic, mixed martial arts fanatic and psychedelic adventurer Joe Rogan. Through Joe's career he's interviewed the iconic entrepreneur Elon Musk, the prolific angel investor Naval Ravikant, the singer, songwriter, actor and rapper, Wiz Khalifa, the English comedian, actor, author and activist, Russell Brand, editor-in-chief of the Daily Wire Ben Shapiro, the comedian, actor and producer, Kevin Hart, etc.

Although I can personally say that I cannot possibly understand why anybody would ever want to listen to *The Report, The Clearing, Noble Blood, Moonrise,* and almost anything ever that is not both practical and actionable, it frankly does not matter because the hosts of these shows are making copious amounts of cash because the world is listening to the podcasts that they are consistently producing and sharing with the planet, one download at a time.

Personally, as a general rule I only listen to podcasts that have the ability to help me practically improve an area of my life and those shows include:

Entrepreneurs On Fire podcast hosted by John Lee Dumas

The Tim Ferriss Show podcast hosted by Tim Ferriss

The Potter's Touch podcast hosted by Bishop T.D. Jakes

The ***And the Writer Is...*** podcast hosted by Ross Golan

The ***School of Greatness*** podcast hosted by Lewis Howes

The Joe Rogan Experience hosted by Joe Rogan

Fundamentally and foundationally, I believe that knowledge without application is meaningless.

As you are recording your audio, you must keep in mind that you are creating your content for a specific audience. APPARENTLY, there are people on the planet who just simply cannot stop themselves from listening to a podcast about a serial killer. I don't get it. I don't understand it, but APPARENTLY MANY PEOPLE WANT TO HEAR *The Clearing* podcast. For the record, you couldn't pay me to listen to *The Clearing* podcast because I believe that the mind is what the mind is fed and I don't want to feed my mind a steady diet of stories and facts related to the thought habits and daily action steps taken by a serial killer.

"VISION WITHOUT EXECUTION IS HALLUCINATION."

THOMAS EDISON

(The creator of the modern light bulb, the inventor of recorded audio, the inventor of recorded video, and the founder of General Electric.)

"ACTION IS THE REAL MEASURE OF INTELLIGENCE."

NAPOLEON HILL

(The best-selling author of *Think and Grow Rich* and the personal apprentice of one of the world's wealthiest men of all-time Andrew Carnegie.)

So before you start writing the outlines for your podcast and start recording your podcast I would ask you to ask yourself the following 5 tough questions:

» **Who is your ideal and likely audience?** _____

_____

» **Are men or women most likely to listen to your podcast?** _____

_____

» **Who will not like your podcast (specifically what demographics of people)?**

_____

» How long will your podcast be? Will you record short and quippy podcasts or a long-form podcast like The Tim Ferriss Show where the average show is longer than 1 hour?

_____

» Why would people listen to your podcast versus doing something else or listening to something else in this world of endless entertainment opportunities?

_____

» Now that you have taken time to answer the questions listed above, I would ask you to not start recording until you have thoroughly invested the time needed to determine your REAL MOTIVE for creating, launching and sustainably recording it. Why are you doing this?

_____

Throughout this book, I am going to teach you the specific technical steps that you must take in order to record a high-quality podcast. However, it is up to you to decide whether it is worth starting a podcast.

Yes, you could invest the time to learn to master the playing of the guitar, but is it worth it to you? Yes, you could compete in an amateur bodybuilding challenge, but is it worth it to you?

Starting a podcast is not expensive, complicated, or even hard. However the commitment, diligence, attention to detail, the focus, and the willingness to prepare that it takes to grow a podcast audience is rare.

So if you know deep down inside that you are simply going to read this book and implement the content found within it for 1 week in a row, it's a healthier choice to simply save yourself and your family the time, the embarrassment, and cognitive dissonance and to not start the podcast that you are not committed to at all. Are you committed (yes or no)? _____.

SINCE YOU ARE COMMITTED, LET'S CONTINUE

Moving forward, I will assume that you are committed to your quest, your cause, and your craft. Thus, I will be very candid with you and I will share with you that when you start your podcast you will not be good. In fact, when I started my podcast I, too, sounded bad. Almost every time that I have forced a client of mine to sit down and actually listen to their own podcast they are generally un-amazed, embarrassed, or taken-aback by how they actually sound. I'm not sure what they think they sounded like before recording their own podcast, but most people are generally not happy with the way that they sound when they first hear the early recordings of their podcast. However, you should not be discouraged by how you sound when you first launch your podcast.

When a baby first learns to walk it stumbles and unless the baby's parents are awful humans, most parents typically don't mock their baby for being unable to walk. Babies take time to learn to walk, and you my friend will also take time to learn how to walk the way you want to sound via recorded sound.

As you record each and every one of your podcasts, it is absolutely vital that you take the time out of your schedule to listen to your own content so that YOU can hear YOURSELF and both, how good and how bad you actually sound.

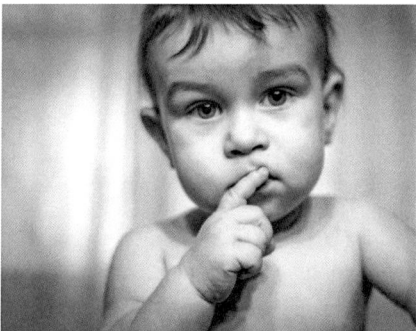

I know that for me personally, when I first began recording our podcast and AM talk radio show, I never really had a problem with my tone of voice because I had worked as a disc jockey, entertainer, and professional business trainer at over 2,000 live events. But, I did immediately recognize that my pacing, energy, commitment to certain character voices, the way I asked questions, and used pauses could all be greatly improved.

When it comes to improving your personal performance, my advice to you is simple, but it will require several hundred hours of your time to implement.

» **Step 1 - Determine who the best podcast in the world is (for your chosen niche).**

_____

» **Step 2 - Record your podcast every day and by that I mean EVERY DAY.**

_____

» **Step 3 - Listen to the best podcast in the world for your chosen niche EVERY DAY.**

_____

» **Step 4 - Ask yourself on a daily basis…"What could you do to improve your podcast to be as good as the best podcast in the world for your chosen niche?"**

_____

» **Step 5 - Write down how you can improve your performance.**

_____

» **Step 6 - Work hard to improve your performance on each and every show. Make it your mission to get 1% better on each and every show that you produce.**

_____

My entire purpose for writing this book is to provide you with the specific

"IDEAS ARE EASY. IMPLEMENTATION IS HARD."

GUY KAWASAKI

(A Thrivetime Show podcast guest, an American marketing specialist author, and Silicon Valley venture capitalist. He was one of the Apple employees originally responsible for marketing their Macintosh computer line in 1984.)

actions steps that you must implement in order to build a successful podcast and to build an audience while starting from the bottom and with an audience of zero. Although most people may think that you are crazy, I do not and I would encourage you to diligently highlight, underline and circle the portions of the book that you want to implement so that you can become a POWERFUL IMPLEMENTER of the SUPER MOVES taught within this book.

As you begin implementing the proven processes and systems found within the pages of this book, please understand that this system works for us and it will work for you.

I remember when we first started *The Thrivetime Show,* although my partner, Doctor Robert Zoellner, and I had achieved tremendous financial success in the world of business, we had no experience in the world of broadcasting or podcasting. Yet today, we now produce and release nine shows per week and receive millions and millions of downloads each year.

Despite our previous lack of knowledge about the world of podcasting as of the writing of this book, our podcast has hit the top of the iTunes charts six times and we've been able to interview some incredible house-hold names on the show thus far, and we are just warming up.

This book was written to help you navigate the technology, the invitation of potential guests, the creation of the outlines and the creation of the buzz around your podcast guests. However, when you first launch your podcast and you are receiving a demoralizing 9 downloads of each podcast, do not lose faith and understand that building a loyal audience is a process and something that is truly the design of a decade and not something that happens overnight.

☺ **FUN FACTS**

Now, known as one of the top earning podcasters on the planet (He brought in $237,193 of income during August of 2017), John Lee Dumas had to start somewhere just like you. Upon leaving the U.S. Army, he enrolled at Roger Williams University School of Law in Bristol, Rhode Island. He then dropped out after only one semester. Dumas took a took a job working in corporate finance for John Hancock in Boston. He then went to New York City to begin working for a technology startup before moving to San Diego in 2009.

In 2011 he moved back to Maine to start a career in commercial real estate. While getting started in this new career he found himself listening to podcasts and he noticed that none of his favorite podcasts provided content daily, and that's when he had his epiphany. He believed that if he could develop a daily podcast focused on interviewing successful entrepreneurs that would resonate with his target audience of entrepreneurs. He came up with the name, Entrepreneurs On Fire and decided to call it EOFire for short.

Daily he continued to grind away after first launching his podcast on September 22, 2012. More and more subscribers began to flock to his podcast as a result of his consistent passionate delivery, the daily format, and his ever rising search engine rank (the more podcasts you produce and transcribe, the higher you will rise in Google search results) he then created the book called, *Podcast Launch*, which he published in the Amazon bookstore. In 2013, fueled by the success of both his book and his podcast, he launched the Podcasters' Paradise (which is a community of podcasters).

As a quick word of caution, if you are not committed to your podcast like

"YOU MUST BE THE PIG AT BREAKFAST AND NOT JUST THE CHICKEN. THE PIG GIVES HIS LIFE FOR THE BREAKFAST, WHEREAS THE CHICKEN JUST LAYS AN EGG. YOU MUST GIVE YOUR LIFE TO YOUR BUSINESS."

DR. ZOELLNER

(The CEO of Thrive15.com and the founder / partner of several multi-million dollar businesses - Z66AA.com. DrZoellner.com, BankRegent.com, RockinZRanchOK. com, DrZZZs.com, www.ZoellnerMedicalGroup.com/, etc...)

John Lee Dumas you shouldn't start a podcast. If you are not willing to commit to diligently working to achieve your dream for 5 years without any positive feedback, don't start a business.

IT IS POSSIBLE FOR YOU TO ACHIEVE MASSIVE SUCCESS, NO MATTER WHAT YOUR BACKGROUND IS:

President Abraham Lincoln, the 16th President of the United States experienced many setbacks throughout his career as you will see in the timeline provided on the next page. However, despite not having earned the respect of his peers by obtaining a college degree, he went on to become a lawyer and president of the United States.

Because Abraham passionately believed in on-going, self-taught education, he never stopped learning (up until the time of his assassination).

PRESIDENT ABRAHAM LINCOLN'S TIMELINE FOR SUCCESS AND STRUGGLE:

1809 - Abraham was born in the state of Kentucky and he grew up in a poor family that lived out on the frontier.

1832 - Age 23 - Abraham lost his job and was defeated for state legislature.

1833 - Age 24 - Abraham officially failed in business.

1835 - Age 26 - Abraham Lincoln's girlfriend died.

1838 - Age 29 - Abraham Lincoln was nominated for the Illinois House Speaker by the Whig caucus, but he ultimately was defeated.

1843 - Age 34 - Abraham lost his attempt to be nominated for Congress.

1846 - Age 37 - Abraham was elected to Congress.

1848 - Age 39 - Abraham lost his renomination for Congress.

1849 - Age 40 - Abraham was rejected in his attempts to become land officer.

1849 - Age 40 - Frustrated with how government works, Abraham left his political pursuits and the world of government to restart his law practice. However, because he was so angered by the Democrats who pushed forth legislation which made it legal for the prairie land regions to own slaves he decided to hop back into politics in 1854.

1854 - Age 45 - Abraham was defeated for U.S. Senate.

1856 - Age 47 - Abraham was defeated for his nomination for Vice President.

1858 - Age 49 - Abraham was defeated for U.S. Senate.

1860 - Age 51 - Abraham Lincoln was elected President of the United States.

DO I QUALIFY TO LAUNCH AND HOST MY OWN PODCAST?

WHAT IF I DON'T HAVE A DEGREE?

SEE A LIST OF PEOPLE THAT DON'T HAVE A DEGREE THAT WENT ON TO DO BIG THINGS.

THE HIGH SCHOOL DROPOUT WHO STARTED ONE OF AMERICA'S LARGEST BANKS.

After dropping out of high school, Armadeo Giannini founded the Bank of Italy, which later became the bank that we all know now as the Bank of America. Giannini is credited as being the father and the creator of many of today's modern banking practices including being one of the first bankers to intentionally offer banking services to middle-class and struggling Americans and not just to upper-class people. At the young age of just 34, Giannini started the Bank of Italy in San Francisco. His original bank was located in a former bar (saloon) and was created to be an institution for the every man and for the "little fellow." He created his bank to be a bank for the hardworking immigrants whom other banks at the time simply would not serve. His philosophy was to judge his clients not based upon their wealth, but based upon their character.

THE GRADE SCHOOL DROPOUT WHO BECAME ONE OF THE WORLD'S WEALTHIEST MEN.

To help his family pay the bills, Andrew Carnegie began working full-time to help support his family for just $1.20 per week or $36.25 in today's money (adjusted for inflation as of 2018) at just the age of 13. Despite being an elementary school dropout, this man went on to become the world's wealthiest man during his lifetime.

THE MAN WITH NO FORMAL EDUCATION WHO WENT ON TO BECOME A U.S. PRESIDENT

Andrew Jackson went on to become an attorney, a U.S. president, a general, a judge, and a congressman despite being home-schooled and having no formal education at all.

THE HIGH SCHOOL DROPOUT WHO BECAME THE MULTI-MILLION DOLLAR PRINCESS OF PRETZELS.

Anne Beiler went on to start Auntie Anne's Pretzels and to become a millionaire, despite having dropped out of high school. I bet she's disappointed she missed out on the once-in-a-lifetime experiences that so many college graduates with $100,000 of debt enjoyed.

THE MAN WHO BECAME THE WORLD'S BEST PHOTOGRAPHER WITH NO FORMAL TRAINING

I don't know if you are into world-famous photographers or not, but if you are, you know that Ansel Adams became arguably the world's best photographer despite not graduating from a college of liberal arts. I wonder how he even knew to take the lens cap off of his camera without a college degree.

THE MAN BECAME THE CEO OF PARAMOUNT AND THE CEO OF FOX WITHOUT A COLLEGE DEGREE.

Barry Diller became a billionaire and a media mogul known for starting the Fox Broadcasting Company, yet he did not have a college degree. Diller got his start working in the mailroom of the Williams Morris Agency after dropping out of UCLA. However, to his credit, Barry did survive three weeks of the non-practical education provided at UCLA. Throughout Diller's career at Paramount while serving as the Chairman and Chief Executive Officer, the company released hit TV shows including: *Cheers, Taxi, Laverne and Shirley* and hit films including: *Grease, Saturday Night Fever, Raiders of the Lost Ark, Indiana Jones and the Temple of Doom, Beverly Hills Cop,* etc. Diller is worth an estimated $3.3 billion.

THE PROLIFIC POLYMATH WHO SAVED AMERICA, HELPED TO CREATE IT AND WHO BECAME A WORLD-RENOWNED INVENTOR DESPITE NOT EARNING A COLLEGE DEGREE

Benjamin Franklin invented the Franklin stove, lightning rods, bifocals, and other assorted inventions while also serving as one of the "Founding Fathers" of the United States, and yet he did not have a college degree.

THE BILLIONAIRE BOSS WHO FOUNDED CLEAR CHANNEL MEDIA WITHOUT A COLLEGE DEGREE

Billy Joe (Red) McCombs became a billionaire, but did he have a degree? No. And that is exactly why he doesn't get invited to any of those fancy alumni gatherings, which he would be too busy to attend anyway because he's off counting his money. Seriously, if he started counting the billions of dollars he made by founding Clear Channel Communications, he might never finish...

BILLIONAIRE BILLY, THE MAN WHO STARTED THAT LITTLE COMPANY CALLED MICROSOFT WITHOUT A COLLEGE DEGREE

You may not be familiar with his name (if you've been living in a cave for the past 20 years), but Bill Gates successfully co-founded the little company called Microsoft without a college degree. Although Bill does have a net worth of $103.8 billion as of the time of the writing of this book (2019), I'm sure that missing out on the "college experience" is holding him back in some capacity. What do you think?

THE POWERFUL PERFUME POWERHOUSE WHO DOES NOT HAVE A COLLEGE DEGREE

Coco Chanel may have a perfume that bears her name, but YOU AND I CANNOT POSSIBLY BE IMPRESSED WITH HER because she doesn't have a degree.

THE MAN WITHOUT A DEGREE WHO STARTED HIS CHICKEN EMPIRE BASED IN KENTUCKY

Colonel Harlan Sanders dropped out of seventh grade and went to live and work on a farm that was located near his home. At just the age of 13, he left his home and secured a job painting horse carriages in the growing city of Indianapolis. At the age of 14 he decided to move to Indiana to become a farmhand. At the age of 65 Sanders decided to franchise his secret recipe for that "Kentucky Fried Chicken" for the first time to an entrepreneur by the name of Pete Harman who was from South Salt Lake, Utah and who was already the operator of one of South Salt Lake's largest restaurants. Within just a year of selling the product he was able to triple his sales with 75% of the sales increasing as a result of sales from Colonel Sanders' fried chicken. At the age of 73 and in the year of 1964, Colonel Sanders sold his company for $2 million to a partnership of businessmen based in Kentucky which was headed by John Y. Brown, Jr. who was at the time 29 years old and who would later become the governor of Kentucky. A fun fact, $2,000,000 in 1964 would be worth $16,525,354.85 as of 2019.

THE WONDERFUL WENDY'S FOUNDER WHO STRUCK GOLD DESPITE NOT HAVING A DEGREE

Dave Thomas was born in 1932, and yet as of 1969, be believed that he could still not find a good hamburger in Columbus, Ohio. Despite not having a college degree at the age of 37, Dave Thomas started his now legendary *Wendy's* franchise. Dave decided to name his business after his daughter Wendy who was at the time just 8 years old. Due to Dave's relentless focus on producing the best possible food, the franchise business quickly grew into a 1,000 + store franchise. In 1989 (the same year that the San Francisco Giants competed against the Oakland Athletics in the Major League Baseball World Series), Dave Thomas decided to become the company's spokesperson at the ripe old age of 57.

THE MAN WHO CO-FOUNDED DREAMWORKS WITH STEVEN SPIELBERG AND JEFFREY KATZENBERG DESPITE NOT HAVING A COLLEGE DEGREE

David Geffen dropped out of college after completing only one year and went on to co-found one of the most successful Hollywood content creation companies of all time. Geffen told Forbes that reading *Hollywood Rajah* (the life story of the movie mogul Louis B. Mayer) made him think to himself, "I looked at these moguls and the world they created and figured it would be a fun way to make a living." As of the writing of this book, David is worth an estimated $8.4 billion.

THE MAN WHOSE PASSION WAS MORE THAN JUST A HOBBY

Despite not having a college degree, David Green has gone on to build a billion-dollar business. David took his initial $600 investment and famously turned that into the Hobby Lobby empire that is today worth $7.4 billion.

DAVID KARP

Without a college degree, David Karp went on to form and found the short-form blogging platform that went by the name of Tumblr. At the age of 15, David decided to dropout of school. In 2007, and at the age of 21 David launched Tumblr in February of 2007. In May of 2013, and when David was just 27 years old, Yahoo! announced that it was acquiring Tumblr for $1.1 billion.

DAVID ORECK

David Oreck is a college dropout and the multi-millionaire founder of the Oreck vacuum company. The company now employs more than 1,500 people at its retail stores and is headquartered in Nashville, Tennessee.

DEBBI FIELDS

Despite not having a college degree, little Debbie created a HUGE COMPANY by the name of Mrs. Fields Chocolate Chippery.

DEWITT WALLACE

DeWitt founded *Reader's Digest* despite not having earned his college degree.

NOTE: It took me about three weeks to alphabetize this list of college dropouts because I don't have a degree.

DUSTIN MOSKOVITZ

Dustin is credited as being one of the founders of that little company called Facebook that only moms, dads, cousins, kids, adults, and humans use. I bet he wishes he had stayed in school at Harvard to earn that college degree.

FRANK LLOYD WRIGHT

Frank became one of the most prolific architects of all-time in the history of humanity, but yet never earned the respect of some academics because he never earned that college degree.

FREDERICK HENRY ROYCE

The man who created one of the most precision focused engine and automotive manufacturing companies of all time (Rolls-Royce) dropped out of elementary school and never earned a college degree.

GEORGE EASTMAN

George was the man who developed a worldwide brand and was known as being the pioneer of popular photography and motion pictures despite not having a college degree.

H. WAYNE HUIZENGA

Wayne is the man who created Waste Management Inc., AutoNation, and was famous for being a co-owner of Blockbuster Video. Throughout his career, Wayne has also been a co-owner of the National Football League's Miami Dolphins, the Florida Panthers of the National Hockey League, and the Florida Marlins of the Major Baseball League. Good thing that all of these business ventures worked out for him because he doesn't even have a college degree to fall back on.

HENRY FORD

The legendary founder of the Ford Motor Company transformed the way American automobiles are produced, yet because he doesn't have a college degree, you can be sure that his father-in-law never respected him.

HENRY J. KAISER

Henry Kaiser never graduated from high school. He never earned a diploma, yet he became the founder of Kaiser Aluminum. He should be thankful that his path to riches worked out because if not he would have never been allowed to become a pharmaceutical sales rep without a degree.

HYMAN GOLDEN

Hyman Golden is the college-degree-free founder of the billion dollar Snapple Beverage Company that was purchased by Quaker Oats Company for $1.7 billion.

INGVAR KAMPRAD

The billionaire founder of IKEA, Ingvar Kamprad, did not earn a college degree and he's dyslexic.

ISAAC MERRIT SINGER

The inventor of the famous Singer sewing machine dropped right out of high school since he was spending every second of his time sewing. We are SEW very sorry for him that he did not earn a college degree.

JACK CRAWFORD TAYLOR

Jack Crawford Taylor served honorably in World War II as a fighter pilot for the Navy and he started the car rental company Enterprise all without a college degree.

JAMES CAMERON

James Cameron wrote and directed *Aliens* (1986) and *The Abyss* (1989). He then (without a college degree) solidified his reputation as being one of the best directors on the planet when he directed, *Terminator 2: Judgment Day* (1991). In 1994, James Cameron then created two of the biggest blockbuster movies of all-time, *Titanic* (1997) and *Avatar* (2009). Despite not having a college degree, James is now worth a reported $700+ million.

JAY VAN ANDEL

He may be a billionaire and co-founder of Amway but that is not impressive because he doesn't have a degree. I bet you he wakes up thinking everyday, "I really screwed up my life and missed out on learning how to learn when I chose to not earn a college degree."

JERRY YANG

Jerry Yang dropped out of Stanford's PhD program to create Yahoo! Without that PhD I bet he doesn't even know where to store his $2.6 billion, but I bet that YOU and I could help him put that money to good use somewhere. Jerry, if you are reading this and looking for creative places to store your money, our office is currently located at 1100 Riverwalk Terrace Suite #100 - Jenks, OK 74037 as of the time of the writing of this book. If you would prefer to email us, you can email info@ThriveTimeShow.com.

JIMMY DEAN

Jimmy Dean, the mogul of meat and founder of the multi-million dollar Jimmy Dean Empire dropped out of high school at the age of 16, yet still managed to build his business.

JIMMY IOVINE

Jimmy was the son of a longshoreman who started out working as a secretary. But at the age of only 19, his drive had become his mission in life. He was absolutely obsessed with making records, so he started to work as a studio professional in 1972 when one of his friends got him a job as a janitor at a record studio. Not long after that, he found himself working with John Lennon, Bruce Springsteen, and other great artists. In 1973, he got a full-time job at the Record Plant recording studio in New York where he worked on Meat Loaf's *Bat Out of Hell* album and Springsteen's Born to Run. He helped produce albums that have sold over 250 million copies. In 2006 he teamed up with Dr. Dre to found Beats Electronics. Beats was purchased by Apple for a mere $3 billion dollars in May of 2014. Since he doesn't have a degree, we can only hope he eventually will go on to become successful.

JOHN D. ROCKEFELLER SR.

John Rockefeller became the world's wealthiest man after dropping out of high school to support his single mom and family at just the young age of 16. John D. Rockefeller went on to fund and create America's National Parks System, countless uses for fossil fuels and the largest net worth in modern American history. However, John D. Rockefeller never did earn a college degree, but he did found two universities including: The University of Chicago and Rockefeller University.

JOHN MACKEY

The millennial hub and house of incredible organic food, Whole Foods Market, may have been founded by this man and the Whole Foods movement may have swept the nation, but he never did earn that college degree although he did enroll and drop out of six different colleges.

JOHN PAUL DEJORIA

John Paul DeJoria is worth an estimated $2.6 billion. Which, just goes to show that the old pro-college adage was right that reads, "College may be a waste of time, but earning $2.6 billion is not."

JOYCE C. HALL

Joyce C. Hall started that little company named Hallmark without a degree. I'm sure he spent all of his life apologizing to his family and friends for disgracing them by not getting a college degree.

KEMMONS WILSON

After dropping out of high school, Kemmons started the Holiday Inn. But the real question is what degree did he get? And what clothing did he wear in his 20s and on his days off since he hadn't had been able to spend 4 years of his life stuffing his closet full of college branded sportswear? And how did he ever make friends, because he didn't get a chance to meet his life-long friends on a college campus?

KEVIN ROSE

After dropping out of college, Kevin Rose started Digg.com which was later sold. Kevin then invested in Twitter, Foursquare, Square, Facebook and other leading technology companies. Today, without a degree he is now worth a reported $30 million. But, I bet you he really laments about not having that "college experience" that many people now invest $50,000 per year to get.

KIRK KERKORIAN

Kirk doesn't have a degree, but he was known as the "father of the mega-resort" having built the world's largest hotel in Las Vegas on three different occasions. So, we will give him a pass for not having a college degree after dropping out of school while just in the 8th grade. As of 2008, Kerkorian was worth $3.9 billion.

LARRY ELLISON

Larry has earned a $71.6 billion net worth, despite having dropped out of two different colleges. Larry is most widely known as the co-founder of the Oracle software company.

LEANDRO RIZZUTO

All this guy ever did was spend his time building Conair and nothing else. However, just because he is a billionaire, I bet you he deeply wishes that he had a college team to cheer for like most debt-burdened college graduates.

LESLIE WEXNER

Although my wife buys things from the L Brands (the worldwide retail empire that owns Victoria's Secret, Bath & Body Works, and Limited), I am not by any means endorsing Leslie Wexner's decision to drop out of law school to start a billion dollar brand with the $5,000 that he could have handed over to a college.

MARK ECKO

If you are to be one of the few people who values success then you may find Mark Ecko impressive. If you believe earning a degree is what makes people successful, Mr. Ecko is rather unimpressive. Without earning a college degree Mark was able to build a billion dollar brand and over a $100 million net worth.

MARY KAY ASH

In all reality, Prince should have been writing songs about Mary Kay and the pink Cadillacs instead of songs about "Pink Cashmere," "Purple Rain" and "Raspberry Berets" because Mary Kay Ash was incredible despite not earning a college degree.

MICHAEL DELL

Michael may be the billionaire creator of Dell Computers, but he probably never really feels like a billionaire because he never had the chance to experience college and the drunken parties that come with it. With a net worth of $37.6 billion, I bet you he could now throw a fairly elaborate drunken college party.

MILTON HERSHEY

I always like to say "If you never finish the 4th grade, you will spend all of your life making chocolate." This is exactly what the founder of Hershey's Milk Chocolate ended up doing as the founder of Hershey chocolates after dropping out of fourth grade. Milton sold his first chocolate bars in 1900 and ended up investing in the building of his own company town (Hershey, Pennsylvania).

RACHAEL RAY

Rachel is a genuinely happy person that has a real love for people and food. What really makes so many poor, culinary school graduates mad is that she never had formal culinary training. She may have a Food Network cooking show and be a food industry icon, but can you really trust somebody without a college degree to be an expert? I can. And with an estimated net worth of $60 million, apparently many other people can too, but we're just idiots.

RAY KROC

Ray is a high school dropout, yet he was still able to systemize and franchise the McDonald's hamburger chain. Nobody was more passionate about saying, "Would you like fries?" than Ray Kroc.

RICHARD BRANSON

Richard is the founder of many, many companies, including: Virgin Records, Virgin Mobile, and Virgin Atlantic Airways. In 2007 he made it onto the '100 Most Influential People in the World' list. He was also knighted in 2000 at Buckingham Palace for his entrepreneurial success. But what does he know? He dropped out of school when he was just 16 years old.

RICHARD SCHULZE

You've heard of Best Buy, right? It's kind of a big deal. Well, this guy made that happen all without that magical piece of paper known as a college degree. I wonder what he is going to fall back on if Best Buy doesn't work out for him.

ROB KALIN

Rob flunked out of high school and then he started Etsy using a $50,000 investment and the help of two techies. As of 2010, Etsy was worth $300 million.

RON POPEIL

Ron invented houseware appliances like the Beef Jerky Machine, the Chop-O-matic, and the Showtime Rotisserie & BBQ. But wait, there's more! He did not go to college and today he is worth just over $50 million.

RUSH LIMBAUGH

The conservative talk show host that 50% of America loves to hate on is, a college dropout. With an annual salary of just over $84 million in 2017, I bet you he wished he had a college degree in accounting so that he would know how to account for his more than $500 million net worth.

RUSSELL SIMMONS

Russell is the man behind introducing the world to hip hop music. He is also a bestselling author, the co-founder of Def Jam Recordings, the founder of Phat Farm, the co-producer of hit films like *The Nutty Professor*, a multiple-time *New York Times* best-selling author worth $340 million and yet he doesn't have a college degree to fall back on and I often worry about him. Hang in there Russell. Quick fun fact: I did name our cat "Russell" after Russell Simmons. This is 100% true.

S. DANIEL ABRAHAM

Mr. Abraham started Slim Fast and is now worth a reported $2.1 billion. But, where's that degree? Oh, that's right...he doesn't have a degree in nutrition from the University-of-the-Market-Does-Not-Care-About-Whether-You-Have-a-Degree-or-Not.

SAMUEL TRUETT CATHY

"Truett" Cathy attended high school in Atlanta and later served honorably in the United States Army during World War II, but yet he never found the time to earn a college degree. Truett started his chicken empire by opening his first chicken-focused restaurant, The Dwarf Grill at just the age of 25.

SEAN JOHN COMBS

Sean Combs, Puff Daddy, P. Diddy, and now the man simply known as Diddy is a part owner of Ciroc Vodka and now has an estimated net worth of $855 million. Sean could have earned a college degree if he hadn't been investing all of his time discovering Mary J. Blige, The Notorious B.I.G., Jodeci, and other hip hop and R&B artists that he helped to produce and promote in route to revolutionizing the music industry. He let all of these things prevent him from getting that prestigious college degree. Let's take a quick moment of silence for his lost opportunities...I'm still pausing...

SEAN PARKER

Let's all go back to the good old days of 1999 and convince Mr. Parker to stop creating the first peer-to-peer platform, Napster, and to focus on his studies, after all, his success depends on it. Without earning a college degree, Shawn is now worth a reported $3 billion dollars. Shawn is a fan of autodidacticism (self-education and self-teaching), which I am, too.

SIMON COWELL

In 2019, *Forbes* reported Simon Cowell earned $49 million and he is the man behind *The X-Factor and American Idol*, however, can he truly ever achieve success without a college degree?

STEVE JOBS

The company Steve Jobs co-founded with Steve Wozniak is currently worth over $1 trillion dollars, a rather unimpressive number when you consider how high it could be had he earned a college degree. I secretly think Stanford may have mistakenly thought Steve Jobs had earned a college degree when they asked him to give their commencement address in 2005.

STEVE MADDEN

Worth a reported $120 million, some would question Steve Madden's priorities when they discover that he chose to pursue building his $100 million dollar brand rather than to earn a college degree. Some people just prioritize poorly.

STEVE WOZNIAK

He may be a billionaire and the co-founder of Apple, but with no fancy framed degree, how successful can he really be?

THEODORE WAITT

This guy co-founded a company called Gateway Computers. He sold more computers in the 1990's than anyone else on the planet, to compensate for the fact that he did not have a college degree. I hope that he is able to buy the feelings of happiness that you can only experience on a college campus with his $4.3 billion net worth.

THOMAS EDISON

Good ol' Thom invented many things, including the invention of the modern light bulb and the invention of recorded video and recorded audio. Thomas Edison also founded General Electric, but he didn't invent a way for himself to achieve all of that success while also earning a college degree. I bet Thom found himself constantly thinking, "If I just had a college degree, maybe then I would be able to earn success."

TOM ANDERSON

Worth a reported $60 million, Tom Anderson gave us Myspace, which gave us the hit-producing band OneRepublic, the platform they used to finally gain their first record deal. Yet, Tom did not block out the time needed to form a study group and to earn that degree.

TY WARNER

If Ty had gone to college he would have learned about papyrus, the Mesopotamia River Valley, cuneiform and the like, yet he chose to invest his time and money into the founding of Beanie Babies. I'm so glad he was able to experience a little success without the painfully expensive piece of paper known as a college degree. Ty is now worth a reported $2.7 billion dollars.

VIDAL SASSOON

The creator of Vidal Sassoon and co-founder of Paul Mitchell Systems is now known as one of the most famous and successful hair stylists in history. Vidal has products and salons all over the world. His only regret? Not graduating from college. With a college degree I bet he would have been able to provide better hair products and services with his $200 million net worth.

W. CLEMENT STONE

W. Clement Stone founded the billion-dollar insurance company called Combined Insurance. He then went on to found the *Success Magazine* publication and to write numerous self-help books despite not having a college degree.

WALLY "FAMOUS" AMOS

This man did not graduate from high school and created the cookie empire known as, Famous Amos cookies.

WALT DISNEY

The co-founder of the Walt Disney Company didn't graduate from high school, and yet I think he turned out alright.

WOLFGANG PUCK

Despite having dropped out of high school at the age of 14, Wolfgang has opened up 16 restaurants and 80 bistros en route to building an incredible national brand of products that you can find in your local grocer. Shhh... don't share his story though, because if we do we may just collectively burst the college bubble. We can't go around respecting people like this because it sets a bad example for kids. ot everyone can go on to become a successful entrepreneur, but everyone can incur $100,000 of student loan debt before finding a soul-sucking job doing something they don't like as it relates to their major in exchange for a paycheck.

WYCLEF JEAN

Previous to earning his college degree, Wyclef Jean had achieved massive success as the co-founder of the Fugees, and as a song-writer and producer for the likes of Whitney Houston, Shakira, Santana, etc.

QUICK WORD OF CAUTION:

I am now going to teach YOU the specific steps that you need to take in order to launch and sustain a successful podcast. However, you must implement these systems in order to achieve massive sustainable success. Once you know what to do, it will be on you to implement what you have learned even on the days when you don't feel like it.

THE REAL REASONS WHY THE VAST MAJORITY OF AMERICANS CANNOT GET THINGS DONE AND CAN'T AFFORD TO INVEST IN THEMSELVES OR ANYTHING ELSE:

Where Does the Time and Money of Most People Go?

Gambling is expensive and requires large portions of a person's time - "The average American loses almost $400 per year to gambling." According to one study, 27.1 percent of gamblers who reported spending over 5 percent of their gross family income monthly, also report experiencing serious problems because of their gambling habit, including health problems, high debts, financial issues, or guilt and other negative emotions.

Stealing requires alot of time... spent behind bars. "75% of employees steal from the workplace and most do so repeatedly."- U.S. *Chamber of Commerce* and *CBSNews* - https://www.cbsnews.com/news/employee-theft-are-you-blind-to-it/.

Cheating on your spouse / partner requires living a double life and we all have just 24 hours in a given day - "78 percent of the men interviewed had cheated on their current partner." – 5 *Myths About Cheating* – https://www.washingtonpost.com/opinions/five-myths-about-cheating/2012/02/08/gIQANGdaBR_story.html?noredirect=on&utm_term=.05ab54a87466

GOING TO STARBUCKS EVERY DAY IS EXPENSIVE

According to a recent survey of American workers by *Accounting Principals*, Americans who regularly buy coffee throughout the week spend on average, $1,092 on coffee annually."
- *20 Ways Americans are Blowing their Money USA Today* -
KATHERINE MUNIZ https://www.usatoday.com/story/money/personalfinance/2014/03/24/20-ways-we-blow-our- money/6826633/.

WATCHING TV ALL DAY AND INTERACTING WITH SOCIAL MEDIA IS NOT A SUCCESS TIP

"On average, American adults are watching five hours and four minutes of television per day. The bulk of that — about four and a half hours of it — is live television, which is television watched when originally broadcast. Thirty minutes more comes via DVR." - How Much Do We Love TV? Let Us Count the Ways *New York Times* - JOHN KOBLIN - https://www.nytimes.com/2016/07/01/business/media/nielsen-survey-media-viewing.html.

THE AVERAGE AMERICAN SPENDS OVER 11 HOURS PER DAY CONSUMING MEDIA

"American adults spend over 11 hours per day listening to, watching, reading or generally interacting with media." - TIME FLIES: U.S. ADULTS NOW SPEND NEARLY HALF A DAY INTERACTING WITH MEDIA - https://www.nielsen.com/us/en/insights/article/2018/time-flies-us-adults-now-spend-nearly-half-a-day-interacting-with-media/

YOUR SMARTPHONE IS MAKING YOU DUMB

"Imagine that after a routine medical exam your doctor delivers some devastating news: Since your last checkup, your cognitive performance has plummeted. Your ability to connect with others has eroded. And your memory for everyday events is no longer operating as it once did. As it turns out, there is a cure and it won't cost you a penny. The treatment is simple: All that's required is that you put away your smartphone. Few of us will have this conversation with our doctors. But perhaps we should. Over the last few years, scientists have begun studying the way cell phones affect the human experience. And the early results are alarming." - RON FRIEDMAN PH.D. - *Is Your Smartphone Making You Dumb?* - https://www.psychologytoday.com/blog/glue/201501/is-your-smartphone-making-you-dumb

If you take off for your birthday, your spouse's birthday, your anniversary, the days before and after each national holiday, two weeks for vacation, and when you don't feel good while starting or growing a business (before you make your millions), you will be poor.

Circle the days you took off this past year from sowing seeds, and determine how realistic it is for you to plan on reaping a harvest this year.

The day before New Year's Eve

New Year's Eve

New Year's Day

The day after New Year's Day

The day before Martin Luther King Jr. Day

Martin Luther King, Jr. Day

The day after Martin Luther King Jr. Day

The day before President's Day

President's Day

The day after President's Day

The Thursday before Good Friday

Good Friday

The Saturday before Easter

Easter

The day after Easter

The day before Memorial Day

Memorial Day

The day after Memorial Day

The day before Independence Day

Independence Day

The day after Independence Day

The day before Labor Day

Labor Day

The day after Labor Day

The day before Columbus Day

Columbus Day

The day after Columbus Day

The day before Veterans' Day

Veterans' Day

The day after Veterans' Day

The Monday of the week of Thanksgiving

The Tuesday of the week of Thanksgiving

The Wednesday of the week of Thanksgiving

Thanksgiving

Black Friday

The day before Christmas Eve (Known as Festivus for all of your Seinfeld fans)

Christmas Eve

Christmas Day

The day after Christmas

7 days that you don't feel like coming in because you feel sick

104 weekend days off

"YOU CAN'T GET MUCH DONE IN LIFE IF YOU ONLY WORK ON THE DAYS WHEN YOU FEEL GOOD."

JERRY WEST

Hall of Fame basketball player and legendary NBA executive

"THE DOERS ARE THE MAJOR THINKERS. THE PEOPLE THAT REALLY CREATE THE THINGS THAT CHANGE THIS INDUSTRY ARE BOTH THE THINKER AND DOER IN ONE PERSON."

STEVE JOBS

(The co-founder of Apple, the founder of NeXT and former CEO of PIXAR.)

"Lazy hands make for poverty, but diligent hands bring wealth."

PROVERBS 10:4

"NINETY-NINE PERCENT OF THE FAILURES COME FROM PEOPLE WHO HAVE THE HABIT OF MAKING EXCUSES."

GEORGE WASHINGTON CARVER

(The man born into slavery who went on to help economically liberate African Americans as a result of his obsession with finding crops that would help African Americans to create sustainable income as a result of sustainable farming. His endless study of peanuts and sweet potatoes led to innovative technology and best-practices that financially freed many African Americans for the first time. As a direct result of George Washington Carver's passionate and game-changing research, African Americans were able to farm sustainably and to return the much needed nutrients to their farmland which had been ruined and depleted of minerals from the years of planting cotton on the same piece of land over and over.)

"IF YOU CANNOT SAVE MONEY, THE SEEDS OF GREATNESS ARE NOT IN YOU."

W. CLEMENT STONE

(Best-selling self-help author and the founder of Combined Insurance Company of America which sold accident and health insurance coverage.)

As you are reading this book if you EVER feel as though you don't know how to implement what you are learning I would strongly suggest that you book your tickets to attend our in-person *Thrivetime Show* 2-day interactive business workshops (which are currently the world's highest rated and most reviewed). At www.ThrivetimeShow.com or (as of the time of this book's printing) you can actually attend our workshops for just $37 if you take the time to leave us an objective review on either iTunes or on our Google Maps page. After you have left the review just email us a screenshot of it to info@ThrivetimeShow.com. Because I grew up without money, I know what it's like to not have access to mentors and to people who know the proven path, however my heart is to help you, which is why we also have a scholarship package available for people that need a hand up. What's the scholarship program? The scholarship program is a program designed to make attending a *Thrivetime Show* 2-day interactive business workshop affordable to everybody. Just reach out to us via email at info@ThrivetimeShow.com with your name, email and phone number and a member of our team will reach out to you as soon as possible to figure out a game plan to help make it affordable to attend one of our in-person workshops.

My wife and I turned off our air-conditioning and shared a car while I worked three jobs and she worked two in order to fund the advertising for our first big success DJConnection.com. Holding jobs at Applebee's, Target and DirecTV at the same time was tough. However, it would have been tougher to have been content to live a life of mediocrity because of buying into the excuse that "I couldn't afford to advertise" and to invest in my business. Life is all about trade-offs. What are you willing to trade off to become successful?

»   www.ThriveTimeShow.com/Business-Conferences

"PEOPLE WHO ARE UNABLE TO MOTIVATE THEMSELVES MUST BE CONTENT WITH MEDIOCRITY, NO MATTER HOW IMPRESSIVE THEIR OTHER TALENTS."

ANDREW CARNEGIE

(One of the world's wealthiest people who started working at the age of 13 to help support his family's financial needs.)

"PEOPLE AROUND YOU, CONSTANTLY UNDER THE PULL OF THEIR EMOTIONS, CHANGE THEIR IDEAS BY THE DAY OR BY THE HOUR, DEPENDING ON THEIR MOOD. YOU MUST NEVER ASSUME THAT WHAT PEOPLE SAY OR DO IN A PARTICULAR MOMENT IS A STATEMENT OF THEIR PERMANENT DESIRES."

ROBERT GREENE

(Best-selling author of *Mastery, 48 Laws of Power, and other best-selling titles.*)

"WE NEED TO RE-CREATE BOUNDARIES. WHEN YOU CARRY A DIGITAL GADGET THAT CREATES A VIRTUAL LINK TO THE OFFICE, YOU NEED TO CREATE A VIRTUAL BOUNDARY THAT DIDN'T EXIST BEFORE."

DANIEL GOLEMAN

(His 1995 book, *Emotional Intelligence* was on the *New York Times* best-seller list for a year-and-a-half. He received a Career Achievement Award for journalism from the American Psychological Association.)

"EVERY ADVERSITY, EVERY FAILURE, EVERY HEARTACHE CARRIES WITH IT THE SEED OF AN EQUAL OR GREATER BENEFIT."

NAPOLEON HILL

(The best-selling author of *Think and Grow Rich* and the personal apprentice of the late great Andrew Carnegie.)

☺

FUN FACTS

YOU MUST HAVE BIG OVERWHELMING OPTIMISTIC MOMENTUM (BOOM) CREATING ACTION.

You can't allow yourself to become a start and stopper. People that start and stop things are weak, and they create a momentum of start and stopping all around them everywhere they go as they declare reasons to justify why they quit including, "The economy is terrible." "You just can't find good employees today." "I ran out of time." "It's hard to win when you don't have any momentum." or "It just wasn't meant to be." People who make statements like this are weak-ass. Every successful entrepreneur that I've ever had the pleasure to interview has gone through hell and back to find their path to success. You must refuse right now to quit no matter how challenging the situation is and no matter how long it takes you to make a profit or you should not start.

MICROSOFT TOOK 6 YEARS TO GAIN TRACTION

Bill Gates may have founded Microsoft in 1975, but it took him nearly six years to land his big contract with IBM. Have you ever worked on something without traction for six years? Most people that I have met, or coached tell me that they begin to get depressed when something is challenging and they are not seeing immediate results based upon the efforts that they are putting into something, but yet again I would call this "weak-ass-thinking."

When I started DJConnection.com out of my college dorm room in 1999, I did not know that it would take me until 2004 to become sustainably profitable. In fact, I did not care and to this date I do not care how long it takes to become financially successful at something because I enjoy the process. When I was growing DJConnection.com, I focused on making each and every one of my personal shows and every aspect of the business 1-2% better each week.

I realize that Apple has accomplished massive things over the years and in recent years most would say that Apple has become the world's most valuable company, yet for many years the co-founders of Apple (Steve Jobs and Steve Wozniak) struggled to gain financial success. Remember, Apple was started in 1976, but it really didn't "gain traction" until the creation of the Macintosh in 1984 (8 years later). How would you feel if you had been working on something that didn't show much progress after 8 years? I WOULD FEEL GREAT, because I have trained my mind to no-longer process adversity as a negative thing. In fact I 100% believe that only through struggle can strength be created.

Most people love to talk about the epic success of the social media superstar Gary Vaynerchuck, but most people fail to remember where he started. I would encourage you to do a Youtube search right now for "Episode 1 - Verite WineLibraryTV." Make yourself watch the first episode of Gary's show. Watch all 12 minutes and 19 seconds of it. Notice that it was published on May 17th of 2006 and that it does not look good, but Gary had to start somewhere.

DO YOU HAVE THE TENACITY
NEEDED TO ACHIEVE SUCCESS?

When people think of Gary Vaynerchuck today, they usually think about the *New York Times* best-selling author, the inspiring speaker and the man who has built the powerful marketing company, VaynerMedia.

However, Gary started as a Youtube wine critic on his show called *Wine Library* in an attempt to expand his parent's wine business. Gary was actually born in Babruysk, which at the time was in the communist Soviet Union, but today is actually part of Belarus. Gary's family decided to immigrate to the United States in 1978, when Gary was three years old. Gary and his family lived in a small studio-apartment in Queens, New York with 8 family members. At the young age of 14, Gary began working in his family's retail-wine store. Gary graduated high school from North Hunterdon High School and then went on to earn a bachelor's degree from Mount Ida College in Newton, Massachusetts in 1998.

Armed with a degree and a ridiculous work ethic, Gary returned to his family's business in 1998 to take over running the day-to-day aspects of his father's retail-wine store, based in Springfield, New Jersey. The store was named, Shopper's Discount Liquors. Gary chose to rename the store to be called the Wine Library and then he also decided to focus on increasing the size of the family's business by focusing on online sales. As result of Gary's diligence, he was able to successfully grow the family's business from $3 million per year in sales to over $60 million per year in sales.

In 2006, Gary then launched the Wine Library TV Youtube show to be a DAILY webcast that would discuss, critique and educate viewers about wine. I want to make sure that the word DAILY doesn't get missed or skipped over here. Marinate on that word for a second. "DAILY." "DAILY!! "DAILY!!!"

Unfortunately, throughout my career I have seen so many people who are more talented than myself become people that are known for starting and stopping everything that they put their hand to. If you want to succeed you've got to learn to work until your knuckles bleed. You have to unlearn much of the bulls!@% poor teachers and professors taught you in school and in college. If you want achieve tremendous success you can't take off of work for fall break,

Christmas break, spring break, summer vacation, your birthday, your kids' birthday, every time you have a headache or that your kids are competing in some sporting activity. To achieve massive success you just have to put in the time and grind. Nothing is more pathetic and soul sucking than watching a grown man or woman coming up with excuses for not doing what they know they need to do. When Gary launched his "DAILY" *Wine Library TV* show he actually recorded it on a "daily" basis. When Dr. Z and I launched *The Thrivetime Show* podcast over 2,000 episodes ago, I committed to not stop until we got there. Where is there? What is our goal? Our sole goal is to mentor millions and that is a large part of why our show has gained such a loyal following.

"SUCCESS DOESN'T JUST LAND ON YOUR LAP. YOU HAVE TO WORK, WORK, WORK, WORK, AND WORK SOME MORE."

P. DIDDY

(The hip hop pioneer who discovered, produced and promoted Mary J. Blige, Notorious B.I.G., Jodeci, and countless artists into super stardom as the founder of Bad Boy Records. As a result of his numerous business ventures, Forbes now reports that Sean Combs (P. Diddy) is now worth $855 million.)

Our listeners know that they can count on *The Thrivetime Show* to put out high-quality and practical business training every day of the week. I've recorded shows while sick, I've recorded shows at 2 in the morning on the day of our kids' cheerleading performance. I've had to record 7 shows in one day because I knew that our family was going to be out of town on a vacation.

When our studio completely flooded with 3 feet of water LITERALLY filling our studio, we simply began recording at the *Thrivetime World Headquarters* at 1100 Riverwalk Terrace Suite #100 in Jenks, Oklahoma.

"WHERE I EXCEL IS RIDICULOUS, SICKENING, WORK ETHIC. YOU KNOW, WHILE THE OTHER GUY'S SLEEPING? I'M WORKING."

WILL SMITH

(The rapper, turned TV sitcom actor who is now considered to be one of the most sought after actors on the planet having appeared in such hit movies as *The Pursuit of Happyness, Ali, Men in Black, Hitch*, etc. Forbes estimates that Will Smith and his wife Jada Pinked Smith are now worth an estimated combined $300 milllion.)

"WHEN YOU'RE AROUND ENORMOUSLY SUCCESSFUL PEOPLE YOU REALIZE THEIR SUCCESS ISN'T AN ACCIDENT - IT'S ABOUT WORK."

- RYAN TEDDER

(Ryan is the song-writer, singer, record producer and a producer of NBC's hit TV show, *Songland*. Since graduating from Oral Roberts University and interning at DreamWorks SKG in Nashville, Ryan has gone on to write hit songs for Adele, Beyonce, Charlie Puth, Ed Sheeran, Jennifer Lopez, Kelly Clarkson, Maroon 5, Paul McCartney, Selena Gomez, Taylor Swift, U2, etc. Ryan has also been successful in real estate investment.)

"LAZY HANDS MAKE FOR POVERTY, BUT DILIGENT HANDS BRING WEALTH."

- PROVERBS 10:4

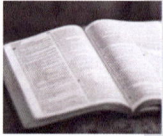

"ON THE SIXTH DAY THEY ARE TO PREPARE WHAT THEY BRING IN, AND THAT IS TO BE TWICE AS MUCH AS THEY GATHER ON THE OTHER DAYS."

- EXODUS 16:5

GOOGLE WAS NOT PROFITABLE FOR ITS' FIRST 3 YEARS

Google was started in 1996, yet as of 1999 very few people had yet to hear of the struggling search engine. Oh, by the way... the company was originally called BackRub. I wonder what would have happened if they had refused to change the name? I wonder if they ever felt stress? Read the true story behind the original Google's first name - http://www. businessinsider.com/the-true-story-behind-googles- first-name-backrub-2015-10

FACEBOOK LOST MONEY DURING ITS' FIRST 3 YEARS OF EXISTENCE (WHY WOULD YOUR STORY BE DIFFERENT?)

When Mark Zuckerberg started this company during his sophomore year at Harvard he originally called it "Facemash" as a way to distract himself from losing a girlfriend. As of 2005, Facebook had posted a yearly net loss of $3.63 million dollars. I wonder if he felt discouraged? I wonder if they ever felt stress?

IT TOOK AMAZON.COM 9 YEARS TO BECOME PROFITABLE

Amazon was founded by Jeff Bezos in 1994 using his life savings and the life savings of his mom and dad's $300,000 retirement. However, in 1999 despite having sales of $1.6 billion, Amazon.com posted a whopping loss of $719 million. In 2003 after having been in business of 9 years, Amazon.com finally posted a profit. How stressed out would you be if you had spent your entire life savings and had built a billion dollar business that just publicly declared a loss of $719 million?

WHEN TED TURNER'S DAD KILLED HIMSELF, IT PROVIDED HIM THE FIRE OF DESIRE HE NEEDED TO SUCCEED

Ted Turner's father literally (this is not an analogy or an exaggeration) told him that he was disappointed by him and then he killed himself. Then Ted went out and leveraged his life savings to buy a small UHF TV 17 station in 1970. Before 1980, fueled by his desire to prove his deceased father wrong, Ted had produced enough revenue and profits to buy the Atlanta Hawks, the Atlanta Braves and Superstation TV 17. I wonder how bad Ted felt when his father blamed him for his suicide? However, he chose to turn his bitterness into betterness.

TESLA'S 10 YEARS SPENT LOSING MONEY

When Elon Musk took over the Tesla company founded by Martin Eberhard and Marc Tarpenning the company was losing money and had been losing money since 2003. Despite Musk's hard work and genius, the company continued to lose money until 2013. I wonder how stressed out Elon Musk must have felt with his life's savings and his reputation on the line?"

YOU CAN DO THIS, BUT DON'T
ALLOW YOURSELF TO GET STUCK

As you are reading this book, if at any point you feel overwhelmed, just book your tickets to attend our next in-person *Thrivetime Show* Workshop. At these workshops we have helped thousands of people just like you to learn the specific steps that they needed to take to turn their dreams into reality. Encourage yourself today by checking out their success stories at:

https://www.ThrivetimeShow.com/testimonials

"NOTHING WAS NEAR THE REALM OF ME HAVING THE SKILLS THAT I NEEDED. SO I LOOKED TO SOMEBODY WHO WAS WHERE I WANTED TO BE."

- JOHN LEE DUMAS

(The Host of the *Entrepreneurs on Fire* Podcast who has been nice enough to feature me on his show twice)

DR. ROBERT ZOELLNER

ThriveTime Show CEO /
Optometrist / Entrepreneur
/ Venture Capitalist

CLAY CLARK

Founder of ThriveTime Show /
U.S. Small Business Administration
Entrepreneur of the Year

BEST amazon.com

Member of the
Forbes
Coaching Council

GAIN TIME&FINANCIAL
FREEDOM

REDUCE YOUR WORKING HOURS, DECREASE COSTS, AND INCREASE TIME FREEDOM + PROFITS!

OUR HOSTS HAVE BEEN SEEN ON

Bloomberg   Forbes   pando   YAHOO!   FSTCOMPANY   BUSINESS INSIDER   Entrepreneur

"SUCCESS IS A CHOICE."

NAPOLEON HILL

(The best-selling author of the book *Think and Grow Rich* which has now sold over 1,000,000 copies.)

CHAPTER ONE

WHAT IS THE POINT OF A PODCAST?

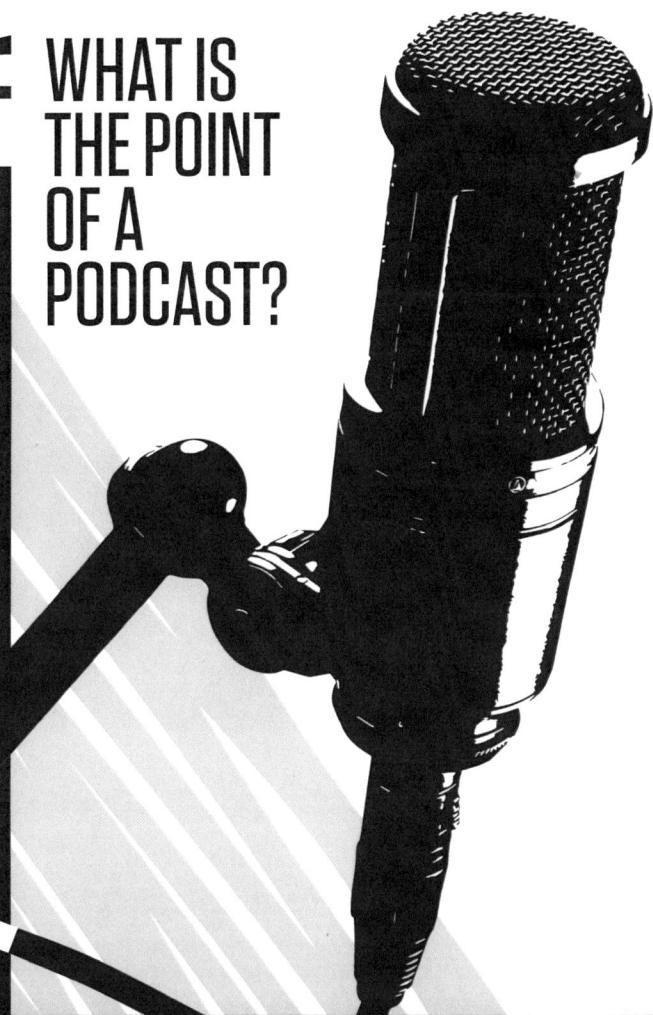

Before you begin obsessing about your outlines, buying equipment, setting up your podcast player, researching your topics, reaching out to and requesting interviews with megawatt, bonafied stars and success stories like Seth Godin, John Maxwell and Wolfgang Puck, it's time for a gut check and to ask yourself the really big questions.

» **Why do you want to record a podcast?**

_____

» **Can you handle both the negative and positive reviews that your podcast will generate?**

_____

» **How long are you willing to put in the work and record your podcast without earning a dime from your show?**

_____

» **Are you truly obsessed with your topic (because if not, you will be competing with somebody who is)?**

_____

CONSISTENCY IS KING

On the planet that we currently live in, there are countless ways that you could get your message out there. You could buy billboards, you could write a book, you could do mass mailers, you could blog, you could record YouTube videos until your head explodes, you could build up a large Instagram following, you could become a prolific Twitter machine, you could become a fabulous Facebook feed dominator, you could produce viral music videos, sensational songs and much, much more. However, because you bought this book, you probably want to start a podcast. Once you start producing content you must understand that your listeners will demand that you ARE MORE CONSISTENT THAN THEY ARE. YES! Listen, if you could faithfully put out one podcast a week, they may listen to only several podcasts, but only a real SICK FREAK (like me) is going to listen to each and every one of your shows. From time to time (if your listeners are mentally healthy) they will cheat on you and listen to another podcast kind of like yours and in your same genre, but over time they will come back and listen to your show if you are truly the best at what you do in your niche.

"BE SO GOOD THEY CAN'T IGNORE YOU."

STEVE MARTIN

(The award-winning actor, filmmaker, writer, comedian and musician behind the hit movies, *The Jerk*, *The Father of the Bride*, *The Father of the Bride II* which eventually caused Steve to be awarded an Emmy, a Grammy and American Comedy awards.)

However, if you stop recording simply because you get your feelings hurt and stop recording because your monthly downloads went down for the first time last month or because your monthly downloads have gone down each month for the last twelve months, the listeners who do come back will immediately quit trusting both YOU and YOUR show as a source of wisdom because YOU are not consistent. If YOU are consistent and stay focused on consistently reaching YOUR core audience, they are going to listen in day after day and week after week to YOUR podcast. After at least 12 months of creating consistent life-changing content you will find that YOUR core ideal and likely buyers and target audience will become addicted to listening to your show in the same way that many Americans are addicted to watching *The Simpsons, Monday Night Football* or *American Idol*.

RESEARCH THE G.O.A.T.S. AND WHAT MAKES THEM GREAT

As soon as possible I would encourage you to OBSESSIVELY LISTEN TO THE SHOWS that are currently dominating your podcast niche and ask yourself why they are considered to be so great. As a business podcaster, I consistently listen to the business and life-optimization podcasts created by John Lee Dumas and his *"Entrepreneurs on Fire"* podcast, Tim Ferriss and his *"The Tim Ferriss Show"*, or Joe Rogan and his *"The Joe Rogan Experience"*. As I listen, I ask myself, "Why do I only have 500,000 downloads per month when these guys have millions of downloads per week?" As I've asked myself that question, I have always found an obvious answer and something better that I could be doing.

YOU MUST COMMIT TO KAIZEN TO WIN AS A PODCASTER OR BUSINESS OWNER

I'm not Japanese so I can't actually claim that I was taught this concept and world-view by my parents, my teachers or college professors, but I can tell you that "Kaizen" is (and has been since the age of 16) my world view even when I didn't know what to call it. The word Kaizen in Japanese means, "Activities that continuously improve all functions."

That's my world-view. I'm obsessed with getting 2% better each and every week. When YOU launch a podcast YOU too must also obsess about getting better with each and every show and podcast. As humans, naturally and by default we typically get 2% worse at everything over time, however it is up to YOU and PEOPLE like YOU to go against the grain and to change that trend.

When you choose to create an audio podcast, YOU must commit to creating a show that you are going to consistently produce for at least the next three years. Don't get discouraged. Having worked with many successful entrepreneurs in the world of podcasting I can tell you that most of them improved and have succeeded simply because they did not quit even when it was clearly not the most practical and logical decision they could have made.

The harsh reality is that with today's affordable technology, nearly anybody can produce ONE GREAT SHOW or maybe even a COUPLE OF GREAT SHOWS. However, how many people can diligently produce and put out GREAT audio content on each and every show? I know that you have it within you to become the most persistent person on the planet and that diligence is always the "Dream Maker." Thus, I will ask you again right now, "Are you willing to produce GREAT audio (or video) content on a daily basis for the next three years of your life without compensation?" If you are, then that means you are committed to greatness and should continue reading this book.

"DESIRE IS THE STARTING POINT OF ALL ACHIEVEMENT, NOT A HOPE, NOT A WISH, BUT A KEEN PULSATING DESIRE WHICH TRANSCENDS EVERYTHING."

NAPOLEON HILL

(Best-selling author of the number one self-help book of all-time, *Think and Grow Rich*.)

REASON 1: MONETIZING YOUR PODCAST

Having met thousands of conference attendees at our *Thrivetime Show* workshops, and while speaking to organizations like Hewlett-Packard, UPS, Maytag University, Farmers Insurance, EXP Realty or by just reading the questions submitted by listeners, it would appear that the vast majority of people who want to launch a podcast have the following character traits in common.

They are obsessed with a particular topic.

They would rather research and discuss this topic than to do almost anything else. They would like to get paid as a result of turning their passion and their podcast into profits.

However, I would challenge you to answer the following questions:

How are you going to get paid from your podcast?

_____

Are you going to sell advertisements on your podcast? (I personally do not like selling advertisements on my show, because I don't want to promote a product or service that I wouldn't personally use.)

_____

Are you going to have a book that you can promote via your podcast?

_____

Are you going to promote products with your podcast?

_____

Are you going to promote services with your podcast?

_____

If you don't make money with your podcast, how are you going to be able to pay your bills? (And don't say GoFundMe)

_____

It is SUPER important for you to know that you must know 10 times more about your subject than your listeners or they will quickly realize that listening to you and your show is a waste of time. So if you aren't willing to invest the time needed to become an expert in your given subject then don't waste your time creating a podcast.

> "IN THE FUTURE, THE GREAT DIVISION WILL BE BETWEEN THOSE WHO HAVE TRAINED THEMSELVES TO HANDLE THESE COMPLEXITIES AND THOSE WHO ARE OVERWHELMED BY THEM -- THOSE WHO CAN ACQUIRE SKILLS AND DISCIPLINE THEIR MINDS AND THOSE WHO ARE IRREVOCABLY DISTRACTED BY ALL THE MEDIA AROUND THEM AND CAN NEVER FOCUS ENOUGH TO LEARN."

ROBERT GREENE

(Best-selling author of *Mastery* and *The 48 Laws of Power, etc.*)

Everyone knows when the presenter is not prepared and does not know their material. You owe it to your audience to prepare and if you are not willing to do that then you need to avoid podcasting."

HOW MOST PODCASTERS GET PAID:

ADVERTISEMENTS

Once you have built a loyal audience, companies that are looking to reach your audience will typically discuss with you about advertising on your show. As an example, if you listen to *The Tim Ferriss Show* titled #367: Eric Schmidt - Lessons from a Trillion-Dollar Coach (available to listen to at https://tim. blog/2019/04/09/eric-schmidt/) you are going to hear commercials for the first 5 minutes of his show) and that is how Tim Ferriss gets paid. However, the amount that you can charge advertisers per download ranges wildly. Some podcasts charge $20 per 1,000 downloads, while other successful podcasters like Tim Ferriss and John Lee Dumas have made up their own pricing structure that best serves them and their listeners. At the end of the day you can charge advertisers what you want to charge them if you have a MASSIVE audience. However, you must have a MASSIVE audience (1,000,000 downloads per week) before you can or should charge any advertisers anything.

AFFILIATE COMMISSIONS

As you become a podcaster with a massive audience you can reach out to the products and services that you ACTUALLY USE and BELIEVE in to see if they would be willing to pay you for recommending their products and services. As an example, on *The Thrivetime Show*, we are constantly asked the following questions:

What camera system do you recommend that I use in my office for both security and training purposes? www.Nest.com

What call recording system and overall phone system do we recommend? www.ClarityVoice.com

What is an affordable franchise that I should buy? www.OXIFresh.com

What attorney would you recommend that I use? www.WintersKing.com

What accounting or bookkeeping software do you recommend that I use? www.QuickBooks.com

What real estate broker do you recommend that I use?

Although we will always only recommend services that we sincerely believe in and use, there is absolutely nothing wrong with getting paid for doing so.

> **"YOU CAN START RIGHT WHERE YOU STAND AND APPLY THE HABIT OF GOING THE EXTRA MILE BY RENDERING MORE SERVICE AND BETTER SERVICE THAN YOU ARE NOW BEING PAID FOR."**
> **NAPOLEON HILL**
> (Best-selling author of the number one self-help book of all-time, *Think and Grow Rich*.)

However, I can candidly tell you that I am not allowed to receive commissions for referring certain products and services and I can tell you that quality companies like Nest.com and Quickbooks.com do not pay us anything for referrals even though we refer them all of the time.

SELLING PRODUCTS AND SERVICES

If you have a product or service that can ACTUALLY SOLVE THE PROBLEMS YOUR LISTENERS HAVE, I WOULD HIGHLY RECOMMEND THAT YOU SHARE OR PROMOTE IT TO YOUR AUDIENCE. However, if the products or services are terrible, don't offer it to anyone, much less your audience.

REASON 2: SEARCH ENGINE OPTIMIZATION DOMINATION (SEO)

Although I've literally written an entire book about search engine optimization, titled *Search Engine Domination: The Proven Plan, Best Practice Processes + Super Moves to Make Millions with Online Marketing*, let me break down 4 keys to dominating search engine results:

1. You must create a Google search engine compliant website (a canonically compliant website) - Learn more at TheBestSEOBook.com

2. You must create a Google mobile optimized website - https://search.google.com/test/mobile-friendly

Carpet Cleaning Quotes | Oxi Fresh
https://www.oxifresh.com › Residential ▾
★★★★☆ Rating: 4.5 - 167,165 reviews
Find a **carpet cleaning quote** from an Oxi Fresh franchise close to your home or business. Get your floors, upholstery, and **carpet cleaning quotes** in a flash

People also ask

How much does carpet cleaning typically cost?   ⌄

How much does it cost for Stanley Steemer?   ⌄

3. You must create more original and relevant HTML (hypertext markup language) text than anybody on the planet in your niche.

4. You must gather more sincere and objective reviews than anybody else on the planet in your niche.

As you record each and every show it is ABSOLUTELY CRITICAL that you invest the time needed to transcribe every show and that you must invest the time needed to weave in your keywords into the transcriptions of each and every show. Transcribing your show into powerful relevant HTML content is by far the most underrated aspect and biproduct of creating a podcast. This just in...people are using Google to search for the products and services that they are looking for. Thus, ranking at the top of the Google search engines for a powerful search term can actually change your life. As my listener's know I am obsessed with the incredible coaching and management skills of Bill Belichick. Thus, when you do a quick Google search for "Bill Belichick's #1 Fan" you will find me at the top of the Google search engine results.

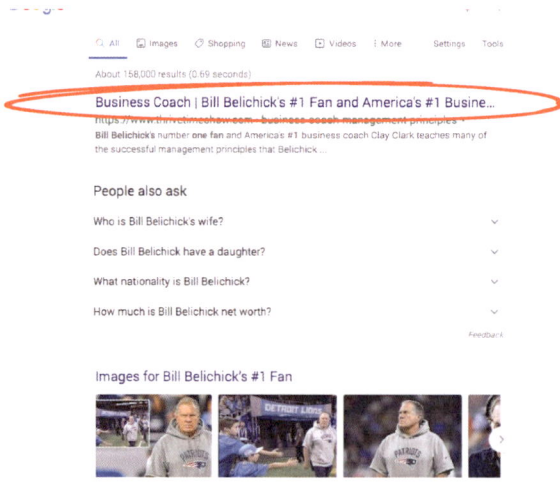

At the time of this writing, when you do a search for "Ross Golan" the hit songwriter for Justin Bieber, Selena, Flo-Rida, etc. and the host of the *And the Writer Is* podcast, you will notice that I rank highly in the search engine results as well.

Each year, I make yet another shameless attempt to book Coach Bill Belichick on *The Thrivetime Show* and each year he rejects me. But the man who has the pleasure of personally shutting me down and telling me respectfully "no" is none other than Coach Belichick's Chief of Staff, Berj Najarian. And thus, you guessed it. If you do a Google search for "Berj Najarian" you will find me again.

FUN FACTS

"AMERICAN ADULTS SPEND OVER 11 HOURS PER DAY LISTENING TO, WATCHING, OR GENERALLY INTERACTING WITH MEDIA."

- Time Flies: U.S. adults now spend nearly half a day interacting with media. 7/31/18 Nelson.com

*REALLY MARINATE ON THE FOLLOWING QUOTE:

"IN THE FUTURE, THE GREAT DIVISION WILL BE BETWEEN THOSE WHO HAVE TRAINED THEMSELVES TO HANDLE THESE COMPLEXITIES AND THOSE WHO ARE OVERWHELMED BY THEM -- THOSE WHO CAN ACQUIRE SKILLS AND DISCIPLINE THEIR MINDS AND THOSE WHO ARE IRREVOCABLY DISTRACTED BY ALL THE MEDIA AROUND THEM AND CAN NEVER FOCUS ENOUGH TO LEARN."

ROBERT GREENE

(The best-selling author of *Mastery*, *48 Laws of Power*, the *Laws of Human Nature*, etc.)

THE ULTIMATE SEO CHECKLIST

In order to dominate the Google search engine results you must implement proven systems and processes. To make your life easier, I have chosen to insert Chapter 2 from our book, *Search Engine Domination: The Proven Plan, Best Practice Processes + Super Moves to Make Millions with Online Marketing*.

In order for you to achieve total SEARCH ENGINE DOMINATION and in order to DRAMATICALLY increase your level of COMPENSATION you must simply check off and complete all of the checklist items on this website evaluation. We humbly refer to this checklist as "The Ultimate Search Engine Domination Checklist."

Google

Google Search   I'm Feeling Lucky

The Ultimate Search Engine DOMINATION Checklist

(and Website Evaluation):

_____ **Host your website with a reliable hosting service**. If your website is hosted with an unreliable hosting service you will rank lower in the search engines. We recommend using GoDaddy.com. Don't host your website with some local, janky hosting provider who lives with his mom in the basement.

_____ **Host your website with the fastest package that you can afford.** Google REALLY CARES about how long it takes for your website to load. Why? Because people get impatient and will quickly move on to another website if your website takes too long to load. On January 17th of 2018, Google formally announced the "Speed Update." Google's plan called for them to slowly roll out the new search engine ranking criteria to give web-developers plenty of time to make their websites load much, much faster. To test the speed of your website visit: https://developers.google.com/speed/pagespeed/insights/ To read more about Google's new speed requirements visit: https://www.forbes.com/sites/jaysondemers/2018/01/29/will-googles-new-page-speed-criteria-affect-your-site/#396634ed6a8f

_____ **Build your website on the WordPress platform.** "WordPress offers the best out-of-the-box search engine optimization imaginable." - Tim Ferriss (Best-selling author of *The 4-Hour Work Week*, *The 4-Hour Body*, *The 4-Hour Chef*, *Tools of Titans,* and *Tribe of Mentors*. He is also an early stage investor in Facebook, Twitter, Evernote, Uber, etc.)

Don't use any other website building platform than WordPress. If you hire coders to custom build your website on PHP or .NET you will end up hating your life as a result of having a website that nobody can update other than the entitled, nefarious employees who now have the ability to hold you hostage. Trust us here. We have personally coached hundreds of clients and every time our coaching clients have a custom built website the business owner at some point has been held hostage by the employee who is the only person who knows how to update the custom built, non-search engine friendly, and ridiculously complicated website. Building your website on WordPress puts the power back in your hands as a business owner because you can update the website yourself if you have to.

PRO TIP: USE WORDPRESS.ORG NOT WORDPRESS.COM

WordPress.org is the open source platform used to power the best SEO compliant websites in the world. WordPress.com is their platform that does not allow for plugins or optimal website optimization.

*Avoid WordPress.com

_____ **Build a mobile-friendly website.** What is a mobile friendly website? Check your website's mobile compliance at: https://search.google.com/test/mobile-friendly. If this link changes in the future just search for "Google mobile compliance test" in the Google search engine and you'll find it.

_____ Install HTTPS encryption onto your website.
HTTPS encryption stands for Hypertext Transfer Protocol
Secure. What does that mean? HTTPS encryption makes
your website more difficult for bad people to hack, thus
making it tougher for very bad people to crash your
website and to use your website as a way to steal the
personal information of your valuable clients and patrons.
Google ranks websites higher who have invested the
additional money needed to add HTTPS encryption to
their website. How many times would you use Google if
every time their search results sent you to websites that
had been hacked into by cyber criminals and internet
hackers?

← → C ⌂ 🔒 https://www.youtube.com

**_____Install the Yoast.com search engine optimization
plugin into your website.** What is Yoast? Yoast SEO is
the best WordPress plugin on the planet when it comes
to search engine optimization. Yoast was built and
designed in a way to make search engine optimization
approachable for everyone, and thus we love Yoast.
Yoast makes it possible for people who are not complete
nerds to proactively manage the search engine
optimization of their website.

DEFINITION MAGICIAN
Plugin - A plugin is a piece of code or software that provides a
variety of functions that you can add to your WordPress website.
Plugins allow you to increase the functional capacity of your website
without having to hire a bunch of nefarious, entitled custom coders
who are typically hard to manage because you do not have any idea
what they are working on or what they are talking about 90% of the
time.

_____Uniquely optimize every meta title tag on every page of your website.
The title tag is simply a hypertext markup language (HTML) element on a website that specifies to search engines what a particular web page is all about. "according to SEOMoz, the best practice for the title tag length is to keep titles under 70 characters." An example would be, "Full Package Media | Dallas Real Estate Photography | 972-885-8823"

Full Package Media | Dallas Real Estate Photography | 972-885-8823
https://fullpackagemedia.com/ ▾
Looking for the best in the business when it comes to **Dallas** Real Estate Photography? You need to

_____Uniquely optimize every meta description on every page of your website. The meta description is simply part of the hypertext markup language (HTML) code that provides a brief summary about a web page. Search engines like Google usually show the meta description in search engine results. Don't make your meta descriptions more than 160 characters in length.

An ample example would be, "Looking for the best in the business when it comes to Dallas Real Estate Photography? You need to call Full Package Media today at 972-885-8823."

Looking for the best in the business when it comes to **Dallas** Real Estate Photography? You need to call **Full Package Media** today at 972-885-8823.
Careers · About Us · Contact Us · Client Login

_____Uniquely optimize the keywords on every page of your website. Meta keywords are a very specific kind of meta tag that will show up in the hypertext markup language (HTML) code on web pages and these will tell the search engines what the web page is really all about. An example of specific keyword optimization would be "Berj Najarian." You may be thinking, who is Berj Najarian?

Berj Najarian serves as the New England Patriots Director of Football and the "Chief of Staff" for the legendary Coach Bill Belichick who has won a total of 8 Super Bowl titles since beginning his coaching career in the National Football League. If someone is searching for "Berj Najarian" there is a high probability that they already know who "Berj Najarian" is and if you want to rank high in the search engines when people are searching for "Berj Najarian" you definitely want to make sure that you have declared your meta keyword phrase as "Berj Najarian."

Quick Note: If at any point while reading this you are beginning to feel overwhelmed just submit your website for an audit and deep dive evaluation and we'll do the heavy lifting for you. You can submit your website to be audited at: www.ThrivetimeShow.com/Website

_____ **Create 1,000 words of original and relevant text (content) per page on your website.** Are we saying that somebody actually has to write, 1,000 original words of original and relevant text for every page of your website? Yes. Isn't there a hack? NO. Can't there be a better way? No.

Can't you just go out and hire a company out of India to use "spinners" to slightly change existing text for you? NO. Can't you just copy content from another website? NO.

You can spend every minute of every day trying to find some blogger or some website experts out there that will tell you that someone on your team doesn't need to invest the time needed to create 1,000 words of both original and relevant content and you will eventually find them and they will be 100% wrong. However, they will gladly take your money.

Google — berj najarian

All | Images | News | Shopping | Videos | More   Settings | Tools

About 6,450 results (0.38 seconds)

META TITLE TAG →

Who is Berj Najarian? | Bill Belichick's Secret Weapon | Thrivetimeshow
https://www.thrivetimeshow.com/.../berj-najarian-the-80-20-rule-the-new-england-pat... ▾
★★★★★ Rating: 4.9 · 2,651 reviews

PERMALINK →

Berj Najarian is Bill Belichick's Chief of Staff he's the human on the planet that has spent the most time with Bill Belichick since he became the New England ...

META DESCRIPTION →

Who is the mysterious Berj Najarian, Bill Belichick's right-hand man ...
https://www.bostonglobe.com/sports/patriots/2019/01/31/...berj-najarian.../story.html
Jan 31, 2019 · Najarian is one of the most powerful figures on the Boston sports landscape, yet most fans have never heard of him.

Images for berj najarian

→ More images for berj najarian   Report images

YOU OR A MEMBER OF YOUR TEAM MUST WRITE 1,000 WORDS OF ORIGINAL AND RELEVANT CONTENT FOR EVERY PAGE OF YOUR WEBSITE.

XML

_____ **Create a Google search engine compliant .XML sitemap on your website.** What is an .XML sitemap? XML stands for Extensible Markup Language. A quality XML sitemap serves as a map of your website which allows the Google search engine to find all of the important pages located within your website. As a website owner unless you hate money, you REALLY WANT GOOGLE to be able to crawl (find, rank, and sort) all of the important pages on your website. Yoast.com has tools that will actually generate Google compliant .XML sitemaps for you. Don't worry, you can do this!

Fun Fact: I had to take Algebra 3 times en route to getting into Oral Roberts University and I was eventually kicked out of college for writing a parody about the school's president "ORU Slim Shady" which you can currently find on YouTube. If I can learn and master search engine optimization you can too!

_____Create a Google search engine compliant HTML sitemap. What's an HTML site map? A hypertext markup language sitemap allows the people who visit your website to easily navigate your website. This sitemap should be located at the bottom of your website and should be labeled as a "Sitemap."

Hiding your sitemap for any reason is a bad idea because Google assumes that if you are hiding your sitemap you are probably trying to hide something. Don't change the background of your website to be the same color as your sitemap's font or do anything tricky here. You want to make sure that your website's sitemap can easily be found at the bottom of your website. See the example below:

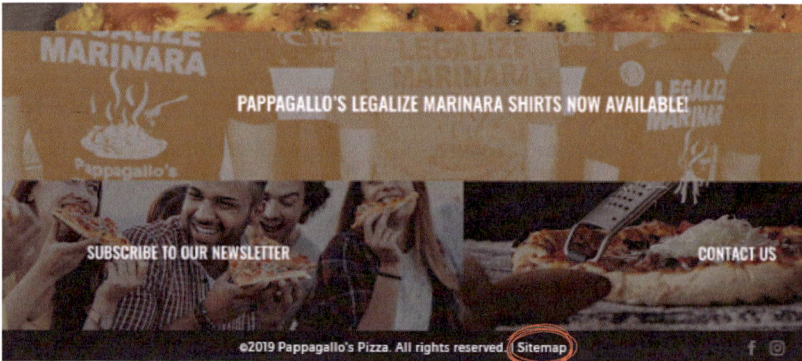

PAPPAGALLO'S LEGALIZE MARINARA SHIRTS NOW AVAILABLE!

SUBSCRIBE TO OUR NEWSLETTER   CONTACT US

©2019 Pappagallo's Pizza. All rights reserved. (Sitemap)

_____Create a clickable phone number. If you ever want to sell something to humans on the planet Earth you must make your contact information easy to find. Thus you want to make your phone number easily available to find at either the top right or at the bottom of your website. When coaching your web-developer, force them to make your phone number a "click-to-call" phone number so that users on your website who are using a mobile phone (almost everyone) can simply click the number to call you.

In our shameless attempt to make this the BEST, MOST HUMBLE and the MOST ACTIONABLE SEARCH ENGINE OPTIMIZATION book of all time we have provided the following real examples from REAL clients just like you who we have really helped to REALLY increase their REAL sales year after year:

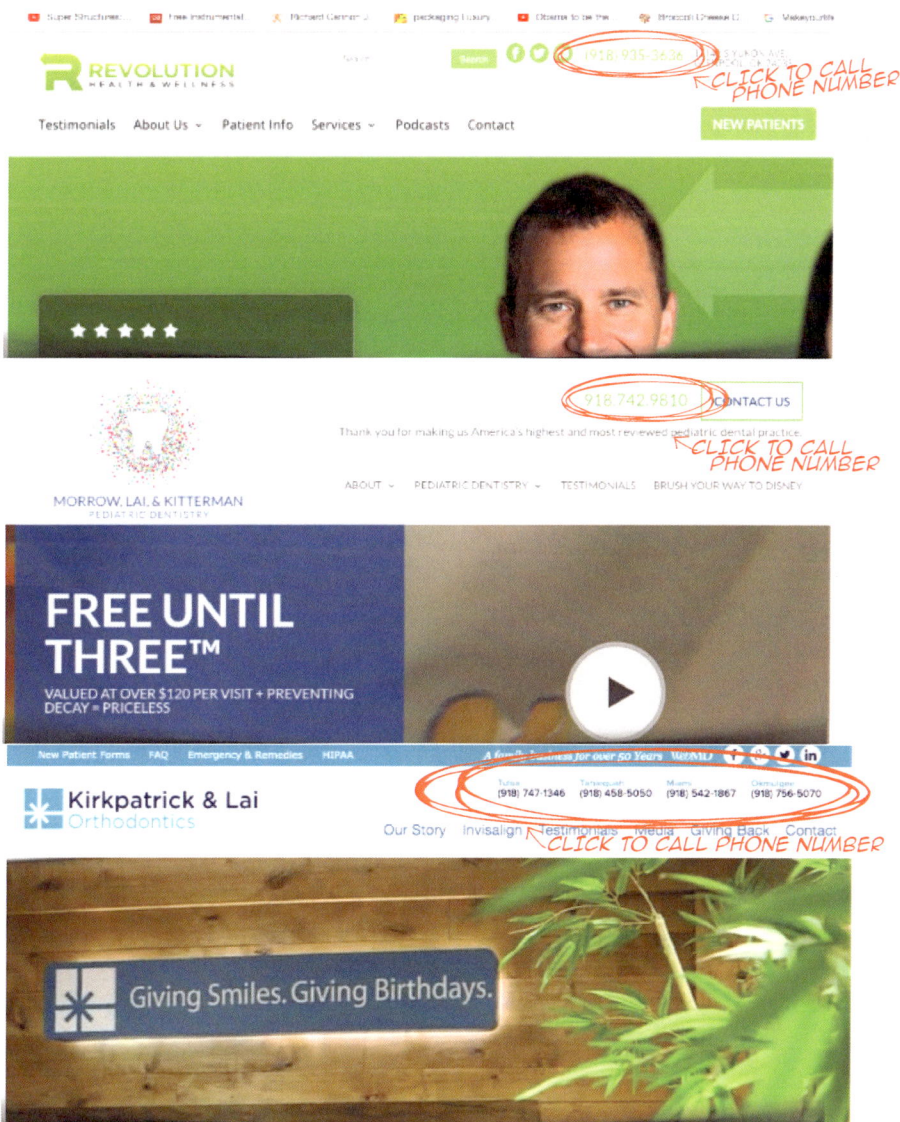

_____Have Social Proof. If you don't hate money and you are not a committed socialist, you will want to include some social proof near the top of your website. What is social proof? "Social proof" is a phrase and a term that was original created by the best-selling author Robert Cialdini in his book, *Influence*. The best social proof examples are:

a. Real testimonials from real current and former clients is super powerful.

b. Media features and appearances on credible media sources like Bloomberg, Fox Business, Entrepreneur.com, Fast Company, etc.

c. Proudly showing that you have earned the highest and most reviews in your local business niche.

d. Celebrity endorsements from celebrities that have earned the trust of your ideal and likely buyers.

e. Listed below is an example that will showcase to you what it looks like to use social proof effectively.

_____**Make the logo return to home.** Allow the logo on your website to serve as your "homepage" button. As of 2019, most people assume that if they click your logo they are going to be taken back to the homepage of your website.

_____**Create original content.** You must create more original and relevant content than anyone else in the world about your specific search engine focus. If you want to come up top in the world for the phrase "organic supplements" you must then create the most original and relevant content on the planet about "organic supplements." If you want to come up top in your city for the phrase "knee pain Tulsa" then you must what? You must create the most original and relevant content on the planet about "knee pain Tulsa."

If you want to come up top in the search engine results for the phrase "America's #1 business coach" then you must create the most original and relevant content on the planet about "America's #1 business coach." Listed below are a few examples of receiving high search rankings due to having the most original, relevant content on the planet about that particular subject.

---

america's #1 business coach   🎤   🔍

All   News   Images   Videos   Maps   More   Settings   Tools

About 5,870,000 results (0.35 seconds)

Business Coach | Bill Belichick's #1 Fan and America's #1 Business ...
https://www.thrivetimeshow.com/the...show/business-coach-management-principles/ ▾
★★★★★ Rating: 99% - 2,651 votes
Bill Belichick's number one fan and **America's #1 business coach** Clay Clark teaches many of the successful management principles that Belichick ...

People also ask

Who is the best business coach in the world?   ⌄

What should I look for in a business coach?   ⌄

Google   Tulsa Tumbling

The Little Gym of SE Tulsa
4.7 ★★★★☆ (14) · Gymnastics center
3.3 mi · 6556 E 91st St · (918) 492-2626
Open · Closes 7:30PM
🌐 Their website mentions **gymnastics classes**

WEBSITE   DIRECTIONS

Twist & Shout Tumbling & Cheer
3.5 ★★★★☆ (8) · Gym
6.2 mi · 4820 S 83rd E Ave · (918) 622-5867
Closed · Opens 5PM
🌐 Their website mentions **tumbling classes**

WEBSITE   DIRECTIONS

≡ More places

Tumbling Tulsa | Tulsa Tumbling Lessons | 918-764-8804
https://justicetumblingco.com/ ▾
If you are looking for the best and highest reviewed **tumbling Tulsa** place, you need to call us at Justice **Tumbling** today and see what makes us better.
Services · About · Schedule · Testimonials

Tulsa Cheerleading | Tumbling Tulsa | Tulsa Tumbling | 918-986-5785
https://tumblesmart.com/ ▾
Tulsa's Most Reviewed **Tumbling** Program. **Tumble** Smart Athletics. Free Evaluation **Lesson**Meet the Owner. **Tumbling Tulsa** Gymnast Stars. Experience the ...
Classes · Facility · About · Testimonials

Google   tulsa knee pain

META TITLE TAG

Tulsa Knee Pain - Revolution Health Tulsa
https://www.revolutionhealth.org/.../tulsa-knee-pain-revolution-health-is-bring-in-a-re... ▾
Find the best treatment for your **Tulsa knee pain** right here in Tulsa. Find out more about Revolution Health by calling at 918-935-3636.

PERMALINK

META DESCRIPTION

Tulsa knee Pain | Revolution Health Oklahoma
https://www.revolutionhealth.org/.../tulsa-knee-pain-find-the-top-and-quickest-result-f... ▾
The best prolotherapy is right here at Revolution Health for **Tulsa knee pain**.

Best Prolotherapy Treatments Tulsa | Tulsa Knee Pain
https://www.revolutionhealth.org/.../tulsa-knee-pain-find-the-best-possible-tulsa-knee-... ▾
Best Care. Best Prolotherapy Treatments for your **tulsa knee Pain**

Non-invasive remedies relieve knee pain without surgery - Tulsa World
https://www.tulsaworld.com/...knee-pain.../article_6bdf681d-d017-554c-9ecc-fae529... ▾
Mar 13, 2019 - Dear Doctor K: I have osteoarthritis of the knee. Are there ways to relieve my **knee pain** without drugs or surgery?

_____Create a "Testimonials," "Case Studies," or a "Success Stories" portion of your website if you want to sell something to humans who were not born yesterday. Most shoppers today have become savvy and are aware of the fact that great companies generate great reviews (and occasionally bad ones) and that bad companies chronically generate bad reviews (and occasionally some good ones). Thus, most people will want to actually see testimonials, case studies or success stories from real clients that have actually worked with your company in the past.

In fact, not having testimonials, case studies, and success stories on your website freaks most people out to the point that they won't even call you or fill out your contact form.

How do we know this? Well, for starters, we are humans who happen to be also consumers say and Forbes tells us that, "Almost 90% of consumers said they read reviews for local businesses. In other words, if you are not investing efforts into online reputation management, then you are missing out on having control of the first impression your business has." - **_Online Reviews and Their Impact On the Bottom Line_** by Matt Bowman - https://www.forbes.com/sites/forbesagencycouncil/2019/01/15/online-reviews-and-their-impact-on-the-bottom-line/#35d3b4955bde

NOTABLE QUOTABLE

"Perfectionism is often an excuse for procrastination."

- PAUL GRAHAM

(The entrepreneur investor, incubator, and coach behind AirBNB, Dropbox, and Reddit.)

 _____**Include a compelling 60-second video / commercial (on the top portion above the fold) on your website** to improve your conversion rate. To provide you with an ample example of clients that we have personally worked with who have used a "website header video" in route to dramatically increasing their sales check out:

VIDEO PLAY BUTTON

_____**Create a "top of the website" call to action** that your ideal and likely buyers will relate to and connect with. You want to make it SUPER EASY for your ideal and likely buyers to call you, to schedule an appointment with you, or for them to do business with you in the most convenient way possible. As an AMPLE EXAMPLE check out EITRLounge.com and OXIFresh.com:

_____ **Create a "No-Brainer" sales offer deal** that is so GOOD, so HOT, and so IRRESISTIBLE that your ideal and likely buyers simply cannot resist the urge to at least try out your services and products out. As an example, we would encourage you to check out the following websites.

"Being top in Google has impacted our business tremendously. Knowing that we're top in Google makes it so much easier for our clients to search and if they use certain keywords that pertain to our business, we're the first ones that come up on that page. We get a lot of phone calls and website traffic. I would suggest every one takes this program seriously."

- MYRON KIRKPATRICK

(Founder of White Glove Auto - WhiteGloveAutoTulsa.com.)

REASON 3: YOUR LISTENERS CAN LISTEN TO THEIR CONTENT 24-7, WHENEVER AND WHEREVER THEY WANT

Your podcast listeners no longer need to synchronize their schedules to match a program's broadcasting schedule. Since the invention of radio, if listeners wanted to listen to iconic personalities like: Glenn Beck, Howard Stern, Steve Harvey, Orson Welles, Paul Harvey or Rush Limbaugh they had to tune in when the broadcaster was actually airing their show live. Now, with podcasts listeners can listen to each and every show that they want when they want as long as they are equipped with a smartphone. As an example, I listen to Ross Golan's *And The Writer Is* nearly every day on the way to and from work and I listen to T.D. Jakes while taking a shower and getting ready for work.

However, many of Golan's shows are longer than my 15 minute commute to work. In fact, his interview with singer / songwriter Matthew Ramsey of Old Dominion is 57 minutes long, yet his interview with Oak Felder the iconic song-writer for Rihanna, Alicia Keys, The Chainsmokers, Nicki Minaj, etc. is 1 hour and 2 minutes long and his interview with the Grammy-winning record producer and the frontman of OneRepublic, Ryan Tedder is 1 hour and 38 minutes long. Now with the podcast format I can stop listening to podcasts once I pull into the workplace and I can resume listening when my workday has concluded and I am on my way home, which is what I do.

FUN FACTS ☺

"The number of podcast listeners has increased sharply this year, according to a new report. More than half the people in the United States have listened to one, and nearly one out of three people listen to at least one podcast every month. Last year, it was more like one in four. "That's the biggest growth we've seen, and we've been covering podcasts since 2006," said Tom Webster, a senior vice president at Edison Research, a company that tracks business trends." - **Jaclyn Peiser** - *Podcast Growth Is Popping in the U.S., Survey Shows* https://www.nytimes.com/2019/03/06/business/media/podcast-growth.html

....

"If your content is boring, no one will listen despite the popularity of the format. - "Some of the biggest U.S. companies are trying to make podcasting the new corporate memo. The problem is getting their employees to pay attention. Verizon's podcast, for instance, averages a few thousand listens and views per episode (despite having a built-in audience of 145,000 workers)

- **Austen Hufford and Patrick McGroarty** - When Corporate America Joins In the Podcast Craze, 'It Gets Boring Fast"

....

"Most podcast consumers listen to most of the podcast episodes they download, and the vast majority listen to at least most of each episode." https://www.edisonresearch.com/the-podcast-consumer-2017/

"While home continues to be the most often named location for podcast listening, the vehicle is a strong second." - https://www.edisonresearch.com/the-podcast-consumer-2017/

"Internet usage worldwide is slowing while smartphone sales have hit a wall. But people are spending more online, talking to smart speakers more, and shelling out more for subscriptions." - Jefferson Graham - Internet usage may be nearing peak, says Mary Meeker in annual tech forecast - *USA Today* - https://www.usatoday.com/story/tech/talkingtech/2018/05/30/internet-usage-growing-slightly-smartphone-sales-have-hit-wall/656078002/v

....

REASON 4:
CONNECT WITH YOUR AUDIENCE

There was a client that we worked with years ago that owned a bridal boutique where future brides could come try on dresses and drink champagne and revel in the excitement of being a bride-to-be. This business owner wanted a way to connect further with the brides that came into her store when she discovered podcasting.

There is a large difference between podcasting and writing a blog. With podcasts, you have the opportunity to more emotionally and empathically connect with your audience in a way that blogging and the written word cannot. You are in the listener's ear talking to them almost as if it were a personal conversation you are sharing. For companies and business owners looking for a way to further connect with their customers, creating a podcast audience is a great way to achieve that.

"ESTABLISH A COMMON DREAM... WANT TO CHANGE THE WORLD? UPSET THE STATUS QUO? THIS TAKES MORE THAN RUN-OF-THE-MILL RELATIONSHIPS. YOU NEED TO MAKE PEOPLE DREAM THE SAME DREAM THAT YOU DO."

GUY KAWASAKI

(*Thrivetime Show* podcast guest and the legendary former key Apple employee turned venture capitalist, best selling author, Chief Evangelist for Canva and Mercedes Benz.)

> "BUILD THE SMALLEST AUDIENCE POSSIBLE. AIM FOR THE SMALLEST POSSIBLE AUDIENCE, NOT THE LARGEST, TO BUILD LONG-TERM VALUE AMONG A TRUSTED, DELIGHTED TRIBE, TO CREATE WORK THAT MATTERS AND STANDS THE TEST OF TIME."

SETH GODIN

(The *Thrivetime Show* guest, iconic entrepreneur, best-selling author of 18 books including *Purple Cow* and the man who in 1998 sold his company Yoyodyne to Yahoo! for $30 million dollars.)

REASON 5: THE BEST THINGS IN LIFE ARE FREE... AND THAT INCLUDES PODCASTS

Podcasts truly are the ultimate "no-brainer" offer. A "no-brainer" offer is a deal that is so good and so enticing that most potential consumers and specifically "potential first-time consumers" simply cannot help themselves, but to give you and your business at least one try. Thus, when you offer your guests FREE unlimited access to potentially life-changing training via your podcast player you NOW have a powerful ability to speak into the lives of your ideal and likely buyers as long as your content is EXCELLENT.

> # "THERE IS NO SHORTCUT TO ACHIEVEMENT. LIFE REQUIRES THOROUGH PREPARATION - VENEER ISN'T WORTH ANYTHING."
> **GEORGE WASHINGTON CARVER**
>
> (The agricultural scientist and inventor who worked tirelessly to encourage the newly freed African American slaves to plant specific alternative crops to prevent their soil from becoming completely depleted of certain essential nutrients.)

REASON 6: YOU WILL QUICKLY BECOME A REAL EXPERT (OR YOU WON'T HAVE ANY REAL LISTENERS)

As you deep dive into the depths of topics you will cover and the guests you will interview you will find yourself truly becoming an expert about your field of study.

The average listener to podcasts has a sound mind that works, and thus they will simply not listen to bad and poorly produced content regardless of how much FREE content that you put out. Thus, if you want to grow your audience over time you are going to have to become an EXPERT as soon as possible. What is an expert?

DEFINITION

An expert is a person who has a comprehensive and authoritative knowledge or a skill in a particular area.

As long as you consistently invest the time needed to prepare high value content for your listeners your audience will keep listening. However, if you fail to prepare for even one of your podcast broadcasts, get ready for your loyal listeners to begin looking elsewhere for both the entertainment and education that they are seeking.

"LUCK IS PREPARATION MEETING OPPORTUNITY."

OPRAH WINFREY

(The media mogul, actress, talk show host, television producer and philanthropist who is now worth a reported $2.6 billion.)

REASON 7:
EXPAND YOUR INFLUENCE

The final and perhaps the most powerful reason for podcasting is that with a well-produced podcast YOU now have the ability to powerfully impact people that are located beyond your geographical area and your current ability to communicate with those people. As of the time I am writing this, I can confidently share with you that our podcast is being faithfully and consistently listened to by entrepreneurs, potential entrepreneurs, podcasters and potential podcasters located all over the world including: Canada, Guam, Australia, Mexico, South Korea, the United States and anywhere where people are allowed to make their own free choices and that capitalism is allowed and encouraged.

Years ago, I helped a physician produce a podcast and it really did take him about 100 podcasts before he had a listener email him saying "I just bought my plane ticket, and I'm excited for my appointment in two days."

This particular Doctor had no idea why patients and people from all across this great United States were willing to schedule appointments with him and his team until he began to ask them how they heard about him. Consistently over-time he began hearing, "Well, you talked about my condition at length on your podcast episode so I wanted to learn more."

As long as you commit right here and now to consistently producing high-quality content over time you will develop a loyal audience.

"REMEMBER THIS: THE GREATS BORE DOWN WHILE THE MEDIOCRE MASS OF HUMANITY STRUGGLES WITH BOREDOM. YOU MUST LASER-FOCUS AND COMMIT TO YOUR CRAFT IF YOU EVER WISH TO MASTER IT WHETHER IT BE PLAYING THE PIANO, LEARNING HOW TO EFFECTIVELY PUBLIC SPEAK, OR LEARNING TO PODCAST."

CLAY CLARK

(Host of the Thrivetime Show and one of the palest males on the planet.)

"IN A CROWDED MARKET PLACE, FITTING IN IS A FAILURE. IN A BUSY MARKETPLACE, NOT STANDING OUT IS THE SAME AS BEING INVISIBLE."

SETH GODIN

(The man who sold his company Yoyodyne to Yahoo! for $30 million.)

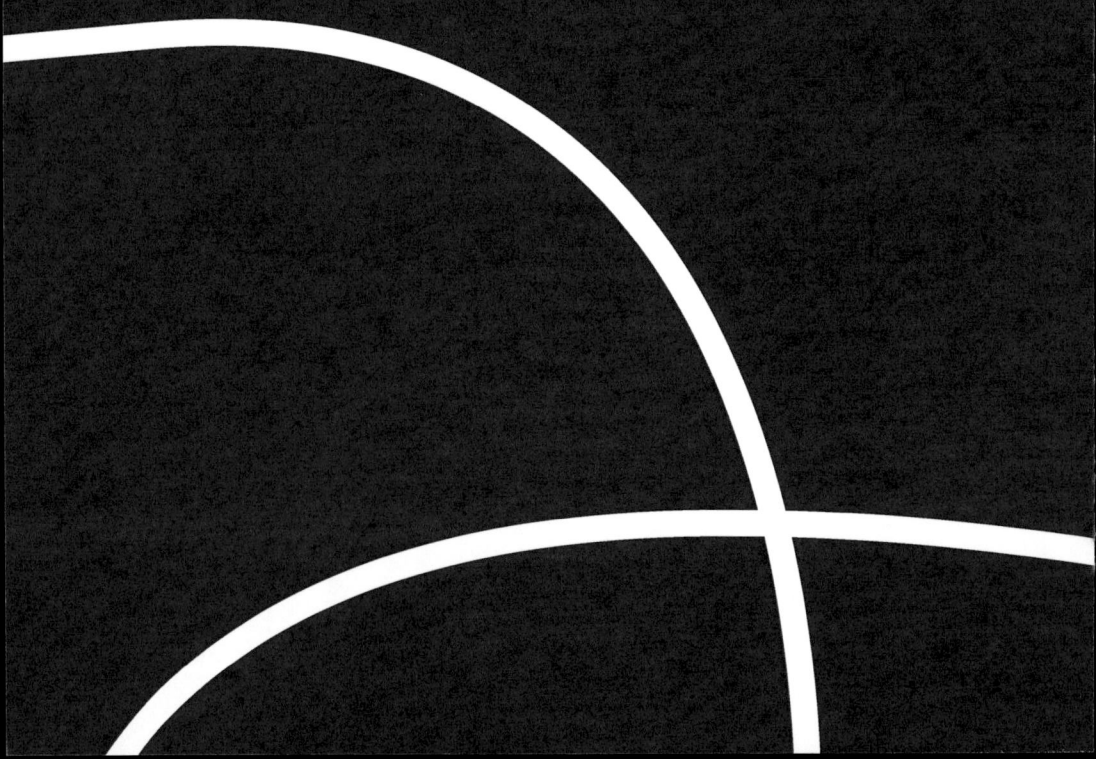

CHAPTER TWO

DETERMINE WHAT YOUR PODCAST IS GOING TO BE ABOUT.

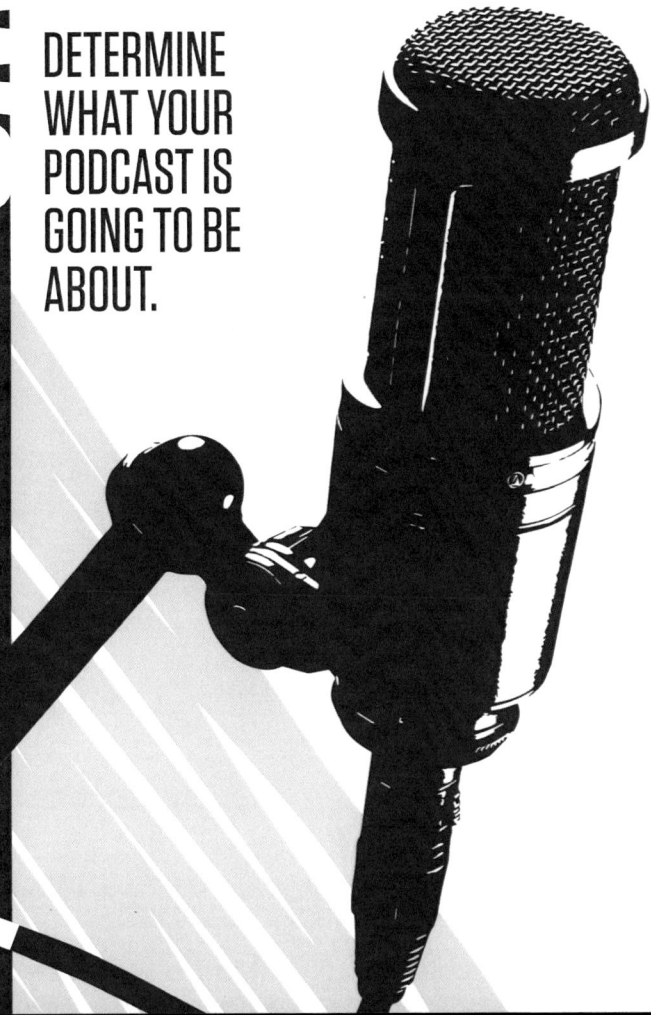

IT IS VITALLY IMPORTANT THAT YOU DETERMINE RIGHT HERE AND NOW WHO YOUR PODCAST IS BEING PRODUCED FOR IF YOU DO TRULY WANT TO EVER DEVELOP A LOYAL LISTENING AUDIENCE.

» What is the ideal age of your ideal and likely listener?

» What is the gender of your ideal and likely listener?

» What is the political leaning of your ideal and likely listener?

» What topic or subject is your ideal and likely listener obsessed with?

» What is your ideal and likely listener passionate about that the average person dislikes?

What subject will you commit to knowing 10x more about than your average listener?

In order to monetize a successful podcast, you must first invest the time needed to define who your ideal and likely listener is and what your podcast is going to be about. You must take the time to define who your ideal and likely buyers are before you launch your show so that you ensure that your podcast will be well received by the people you are trying to reach.

If you truly want to grow your audience, it is important that you provide examples that your listeners can relate to. If your average listener is 40 to 50 years of age they generally do not understand the jargon or the need for jargon being consistently provided by millenials:

"BEST BUYERS BUY MORE, BUY FASTER, AND BUY MORE OFTEN THAN OTHER BUYERS. THESE ARE YOUR IDEAL CLIENTS. HAVE A SPECIAL EFFORT DEDICATED TO JUST THE DREAM CLIENTS."

CHET HOLMES

(Best-selling author of *The Ultimate Sales Machine* and the former business partner of Charlie Munger, Warren Buffett's billionaire partner that no one has ever heard about.)

"YOU MUST UNDERSTAND THE FOLLOWING: IN ORDER TO MASTER A FIELD, YOU MUST LOVE THE SUBJECT AND FEEL A PROFOUND CONNECTION TO IT. YOUR INTEREST MUST TRANSCEND THE FIELD ITSELF AND BORDER ON THE RELIGIOUS."

ROBERT GREENE

(Best-selling author of *Mastery, The 48 Laws of Power, The Laws of Human Nature, etc.*)

"CONDITIONS ARE NEVER PERFECT. 'SOMEDAY' IS A DISEASE THAT WILL TAKE YOUR DREAMS TO THE GRAVE WITH YOU."

TIM FERRISS

(Best-selling author of *The 4-Hour Work Week* and one of the top podcasters on the planet. He has invested or advised in startups such as Facebook, Evernote, Shopify, Reputation.com, and TaskRabbit.)

EXAMPLES OF MILLENNIAL SPEAK 101:

"NETFLIX AND CHILL."

This means that two millennials plan on turning on a movie that they have no intention of watching while fooling around.

"SALTY."

This means that you are in a bad or grumpy mood.

"THIRSTY."

This means that you are chasing or craving something that you want, regardless of whether it is beverage or not.

"TROLLS."

These are people that say or write bad things about people online while hiding behind anonymous account and user names.

"LIT"

This means that something is incredibly awesome.

"MERICA"

This is how many millenials make fun of America's love of itself. When the average person gets super excited about shooting off fireworks or capitalism, many sarcastic millenials will say, "Merica."

When you are recording and producing your podcast it is vitally important that you make sure that your listening audience, "gets the joke" and that they are "feeling you." It is essential that your listeners understand your jokes, your references and the examples that you are providing them on a consistent basis.

DETERMINE THE RIGHT GUESTS AND CO-HOSTS TO FEATURE ON YOUR PODCASTS:

As you get prepared to launch your podcast it will become increasingly important for you to invest the time needed to determine who will be the best fit to help you host your podcast. Personally I prefer to rotate the co-hosts whom I feature on my show, so that my listeners can meet a variety of personalities. However, I would recommend that you would always feature YOURSELF as the host of the show because overtime you will discover that nobody will EVER be as committed as you are when it comes to recording your show on time, every time.

"BELIEVE IN YOURSELF! HAVE FAITH IN YOUR ABILITIES! WITHOUT A HUMBLE BUT REASONABLE CONFIDENCE IN YOUR OWN POWERS YOU CANNOT BE SUCCESSFUL OR HAPPY."

NORMAN VINCENT PEALE

(The American minister and self-help author who became famous for popularizing the concept of positive thinking. Norman became famous internationally as a result of his book, *The Power of Positive Thinking* and as result of leading the Marble Collegiate Church in New York City that was famously attended by President Donald J. Trump.)

"PEOPLE ALWAYS TELL YOU, 'BE HUMBLE. BE HUMBLE.' WHEN WAS THE LAST TIME SOMEONE TOLD YOU TO BE AMAZING? BE GREAT! BE GREAT! BE AWESOME! BE AWESOME!"

KANYE WEST

(The man who has won a total of 21 Grammy Awards, making him one of the most award winning musical artists of all time. Throughout Kanye's career he has produced hit songs for Drake, Justin Bieber, Rihanna, Foxy Brown, Goodie Mob, Jagged Edge, Lil' Kim, Jay-Z, Beyonce, Monica, DMX, Ludacris, Carl Thomas, Maroon 5, D12, Petey Pablo, Jadakiss, Brandy, John Legend, Britney Spears, Mariah Carey, Common, Keyshia Cole, T.I., Diddy, Nas, The Game, etc.)

If you decide to host your own show it can be both FUN and EXCITING, however it can be challenging if you have never previously spoken aloud for over 15 minutes straight without the interaction or interruption from another person. Thus, a super-move that I would encourage you to use is to have a co-host who consistently will join you on your show to keep the energy up and to keep the momentum going throughout your recording. Examples of having a powerful personality to co-host your show include:

» David Letterman was consistently joined by the musician Paul Shaffer and his band on *The Late Show.*

» Ernie Johnson is consistently joined with many supporting characters including Kenny "The Jet" Smith, Shaquille O'Neal and Charles Barkley on the NBA on TNT.

» On *The Thrivetime Show,* Doctor Rober Zoellner often joins me to share both real and raw business principles that blow the minds of the average and "politically correct" listener.

» When you go to ThrivetimeShow.com and listen to my "business yoda" Doctor Zoellner your life will never be the same. He has more talent and mental capacity than I have, yet between the two of us, we have been able to build 15 real multi-million dollar businesses. If we can do it, you can too.

"WORK FOR THAT FEELING THAT YOU HAVE ACCOMPLISHED SOMETHING. DON'T WASTE YOUR TIME ON THIS EARTH WITHOUT MAKING A MARK."

JOE ROGAN

(A stand-up comedian, mixed martial arts commentator, former actor, TV host of *The Joe Rogan Experience* podcast which has featured guests such as: Jim Gaffigan, Jamie Foxx, Dan Aykroyd, Wiz Khalifa, Elon Musk, etc.)

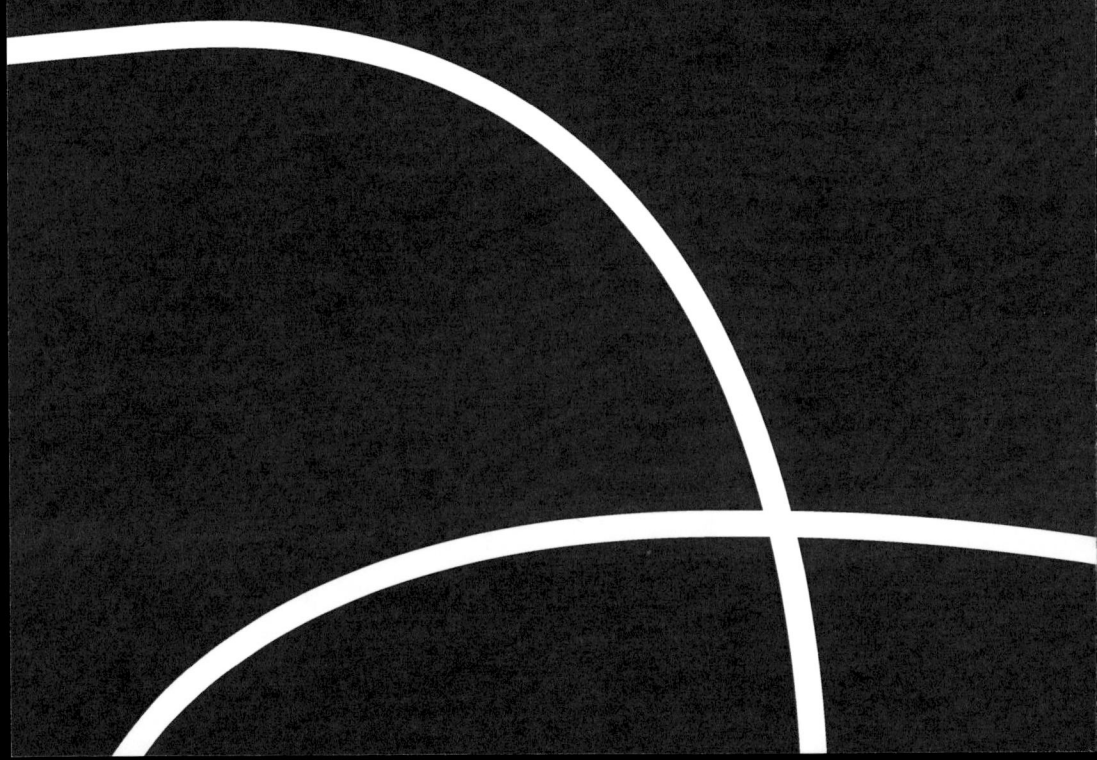

CHAPTER THREE

BECOMING THE CONTENT KING AND WIKIPEDIA OF YOUR INDUSTRY.

YOU MUST CREATE AN OUTLINE FOR EACH AND EVERY PODCAST RECORDING IF YOU WANT THE WORLD TO ACTUALLY TAKE YOU SERIOUSLY

Although it may sound harsh, I want YOU to know that YOU must create an outline for each and every podcast / broadcast that you create. When your audience clearly knows that you are making it up as you go or that you have not adequately prepared for each and every show they will quickly stop listening because they have better things to do. Before each and every show that you produce you are going to need to personally invest in doing your research. You need to invest the time needed to pull notable quotables, to look up cited statistics, to find vetted guests and to think about the personal stories that you are willing (or not willing) to share with your audience before each and every show.

"THE CONVENTIONAL MIND IS PASSIVE – IT CONSUMES INFORMATION AND REGURGITATES IT IN FAMILIAR FORMS. THE DIMENSIONAL MIND IS ACTIVE, TRANSFORMING EVERYTHING IT DIGESTS INTO SOMETHING NEW AND ORIGINAL, CREATING INSTEAD OF CONSUMING."

ROBERT GREENE

(The man who continues to be the best-selling author of *Mastery, The 48 Laws of Power, The 50th Law, etc.*)

My wife and I went to go see Justin Bieber perform live when he came to town, and I might just be a Belieber now. It was one of the most well-rehearsed performances that I've ever seen in the world of music. You can tell that Bieber didn't just start his concert tour by saying "Alright, don't worry about practicing, we're all professionals here. Let's just get out there and hope for the best." My friend, you could tell that he was rehearsing for months in order to get the choreography and music perfect before ever stepping out onto the stage.

In order to create a quality podcast, you need to approach your podcast episode preparation with the level of preparation and attention to detail needed to be taken seriously. Assume the other co-host or guest is not going to be mentally present for your recording and you will need to be able to lead the show recording to the promised land. You and your co-host will always fall to the level of preparation that you require of yourselves.

"WHEN YOU'RE AROUND ENORMOUSLY SUCCESSFUL PEOPLE YOU REALIZE THEIR SUCCESS ISN'T AN ACCIDENT – IT'S ABOUT WORK."

RYAN TEDDER

(Grammy-winning recording artist, singer and songwriter who has written hit songs for U2, Adele, Beyonce, Kelly Clarkson, Jennifer Lopez, Paul McCartney, Taylor Swift, etc.)

WHAT IS THE IDEAL LENGTH OF YOUR PODCASTS?

For the purposes of optimizing your website, I would personally recommend that each and every podcast that you produce is a minimum of 10 minutes long so that when transcribed each and every podcast that you produce is a minimum of 1,000 words long. As you are learning to record your podcasts do not get discouraged. Learning to record is like learning to ride a bike. At first you won't be good. Over time you will develop the basic skills and you will want to wave at your mom as you are riding down the street. Then you are going to lose your balance and hit a curb and rip the skin off of your knee. Then your friends are all going to laugh at you and you are going to attempt to play it off like it doesn't hurt, but in reality it's the worst pain you've ever experienced. But then, over time and as a result of practicing, you get better. Strength can only be gained through struggle, so get started failing TODAY so that you can have big success in the near future."

WHAT DO I SAY? THE FORMAT FOR PODCASTS

Over the years, one of the biggest challenges that I have helped many of our clients work through is teaching them the actual structure of their podcast. It's SUPER important to have your show outline typed out and printed previous to recording so that every guest and podcast host and co-host is literally on the same page at all times.

"THANKFULLY, PERSISTENCE IS A GOOD SUBSTITUTE FOR TALENT."

STEVE MARTIN

(Since the 1980s, having branched away from comedy, Martin has become a successful actor, as well as an author, playwright, pianist, and banjo player, eventually earning Emmy, Grammy, and American Comedy awards, among other honors. In 2004, Comedy Central ranked Martin at sixth place in a list of the 100 greatest stand-up comics. He was awarded an Honorary Academy Award at the Academy's 5th Annual Governors Awards in 2013.)

Before you record each and every podcast it is vitally important that you have the following outline written out that includes:

RAPPORT

Script out your monologue or what you will say during the first 15% of your podcast so that your listeners will like you and will be interested in hearing what you have to say.

NEEDS

Script out the questions you will ask your ideal and likely listeners so that you can truly point out the problems your listeners have that you can solve.

BENEFITS

Script out the solutions that you can provide your ideal and likely buyers. It is imperative that you provide your ideal and likely buyers with proof that what you are saying is true and verifiable.

CALL TO ACTION

Once you have convinced your ideal and likely listeners that what you are saying is 100% factually correct, you must then script out the call to action that you will ask your listeners to take.

ISOLATE OBJECTIONS + SOLVE THEM

Unless you were born yesterday, you will quickly learn that most potential buyers will have the following objections time and time again:

- "I don't have time." - The No-Time Objection
- "I don't have the money." - The No-Money Objection
- "I don't really understand how your services will help me." - The No-Need Objection

"THERE IS ONLY ONE EXCUSE FOR A SPEAKER'S ASKING THE ATTENTION OF HIS AUDIENCE: HE MUST HAVE EITHER TRUTH OR ENTERTAINMENT FOR THEM."

DALE CARNEGIE

(Best-selling author of *The Art of Public Speaking* and *How to Win Friends and Influence People*.)

"YOU DO NOT DESERVE TO TAKE THE AUDIENCE'S TIME IF YOU HAVE NOT INVESTED THE TIME NEEDED TO PREPARE."

CARLTON PEARSON

(Best-selling author and former mega-church Pastor of Higher Dimensions Church.)

"THE WILL TO WIN IS NOT NEARLY AS IMPORTANT AS THE WILL TO PREPARE TO WIN."

VINCE LOMBARDI

(An American football player, coach, and executive in the National Football League [NFL]. He is best known as the head coach of the Green Bay Packers during the 1960s, where he led the team to three straight and five total NFL Championships in seven years, in addition to winning the first two Super Bowls following the 1966 and 1967 NFL seasons.)

Although there are a variety of ways to make your content interesting to listeners, it is important to keep their attention. In order to keep your listeners engaged I would suggest using a combination of the following:

SUPPORTING STATISTICS:

Invest the time needed to find statistics from credible sources that your listeners will acknowledge as being bonafide and true. It's super important that the statistics you cite are not pulled from random blogs or obscure publications that have the habit of making up the statistics.

"THE FIRST PROBLEM OF COMMUNICATION IS GETTING PEOPLE'S ATTENTION."

CHIP HEATH

(*Thrivetime Show* podcast guest and best-selling author of *Made to Stick: Why Some Ideas Survive and Others Die.*)

SUPPORTING STORY

Including supporting stories in each and every podcast is powerful. In fact, using stories is a powerful way to engage the listeners of your show because they always want to know what happens next. If you create a show that involves inviting guests onto your show, you must become an excellent interviewer who has developed the ability to get your guests to share the stories that your audience will find interesting with each and every show.

HOW TO BECOME A WORLD-CLASS INTERVIEWER

Listeners of *The Thrivetime Show* often email in their questions to info@ ThrivetimeShow.com or they will ask me in person at our workshops, "How did you become a good interviewer?"

Although I am getting better each and every year, I would encourage you implement the following 5 pieces of advice to become the best interviewer possible:

» Watch great interviewers like Jimmy Fallon, Ellen Degerees, Oprah and Dave Letterman on his new show, "*My Next Guest Needs No Introduction with David Letterman*" at least once per week.

» As you watch Jimmy Fallon, Ellen, Oprah and Letterman, write down the questions they ask and ask yourself, what makes their questions so good?

» Focus on incorporating self-deprecation into your interviews.

» Make sure that you always bring purposeful and intentional passion to each and every interview?

» Prepare until your head explodes. Spend 4 hours of searching each and every guest per 30 minutes that you plan on interviewing.

"WHAT YOU WANT IN AN INTERVIEW IS FOUR THINGS: YOU WANT SOMEONE WHO CAN EXPLAIN WHAT THEY DO VERY WELL, WHO CAN HAVE A SENSE OF HUMOR AND HOPEFULLY IS SELF- DEPRECATING, WHO HAS A BIT OF A CHIP ON THEIR SHOULDER, AND PASSION. IF YOU HAVE PASSION, A CHIP ON THE SHOULDER, A SENSE OF HUMOR, AND YOU CAN EXPLAIN WHAT YOU DO VERY WELL, IT DOESN'T MATTER IF YOU'RE A PLUMBER OR A SINGER OR A POLITICIAN. IF YOU HAVE THOSE FOUR THINGS, YOU ARE INTERESTING."

LARRY KING

(An American television and radio host, whose work has been recognized with awards including two Peabodys and 10 Cable ACE Awards.)

As a podcast host I would strongly encourage you to ask questions that will prompt your guests to share stories. When your guests begin to share their stories, the listener begins to become immersed in their memories, their descriptions and the details of what they are saying. The Wharton Business College professor and *Thrivetime Show* guest, Jonah Berger, wrote a book titled *Contagious* that specifically explains why and how things catch on (which I highly recommend that you read). Inside his book *Contagious*, he talks about the 6 ways things will catch on, which you can implement as well.

S – Social Currency – People love being in the know and sharing the link that "everyone just has to see!"

T – Triggers – People love listening to Rebecca Black's song *Friday* because it's top of mind every Friday. Find a parade and get in front of it!

E – Emotion – People love clicking on images of puppies, babies, and images they connect with.

P – Public – Most people want to do what most other people are doing. Think about the ALS Ice Bucket Challenge.

P – Practical Value – People love to share tips. Think about The Pioneer Woman who has become the queen and expert of all things DIY (Do it Yourself.)

S – Stories – People love to share stories. When you wrap up your business in artistically cultivated stories, ideas will spread.

"WHEN WE CARE, WE SHARE."

JONAH BERGER

(Best-selling author of *Contagious* and a professor at the Wharton Business School and *Thrivetime Show* guest.)

THE FOLLOWING ARE GREAT QUESTIONS TO ENGAGE YOUR GUESTS:

» Tell me about the time when you _____?

» Describe a situation when you _____?

» Tell me how you felt when _____?

» Looking back...what is the best advice you would give your younger self?

» I know that you've had a ton of success at this point in your career, but I would love to start off at the bottom and the very beginning of your career. What was your life like growing up and where did you grow up?

» When did you first figure out what you wanted to do professionally?

» When did you first feel like you were truly beginning to gain traction with your career?

» I know that you are a serial entrepreneur who has experienced massive success and super low points...walk us through the highest high and the lowest low of your career?

» When you were at the bottom, what did you learn most from this experience?

» Today, I'd love for you to share with the listeners about the kinds of projects that you are working on?

» What are a few of your daily habits that you believe have allowed you to achieve success?

» What mentor has made the biggest impact on your career thus far?

» What has been the biggest adversity that you've had to fight through during your career?

» What advice would you give the younger version of yourself?

» We have found that most successful entrepreneurs tend to have idiosyncrasies that are actually their super powers...what idiosyncrasy do you have?

» What is a message or principle that you wish you could teach everyone?

» What is a principle or concept that you teach people most that VERY FEW people actually ever apply?

» What are a couple of books that you believe that all of our listeners should read and why?

"STORIES CARRY THINGS. A LESSON OR MORAL. INFORMATION OR A TAKE-HOME MESSAGE."

JONAH BERGER

(The man whose name reminds us all of the big whale story, the best-selling author of the book *Contagious*, and a professor at the Wharton Business School.)

INVITING PODCAST GUESTS 101

If you ever want to have featured guests on the show, you are going to have to invite them! When you are first starting out, I have found that you need to send over 900 emails to get one "yes". As your podcast becomes more and more popular and established, this number will drop over time.

"IF YOU GIVE SOMEONE A PRESENT, AND YOU GIVE IT TO THEM IN A TIFFANY BOX, IT'S LIKELY THAT THEY'LL BELIEVE THAT THE GIFT HAS HIGHER PERCEIVED VALUE THAN IF YOU GAVE IT TO THEM IN NO BOX OR A BOX OF LESS PRESTIGE. THAT'S NOT BECAUSE THE RECEIVER OF THE GIFT IS A FOOL. BUT INSTEAD, IT'S BECAUSE WE LIVE IN A CULTURE IN WHICH WE GIFT WRAP EVERYTHING – OUR POLITICIANS, OUR CORPORATE HEADS, OUR MOVIE AND TV STARS, AND EVEN OUR TOILET PAPER. PUBLIC RELATIONS (BRANDING) IS LIKE GIFT WRAPPING."

MICHAEL LEVINE

(My friend and the man who has been the public relations consultant of choice for Nike, Pizza Hut, Charlton Heston, Nancy Kerrigan, Michael Jackson, Prince, etc.)

The more money you make, the more you will begin to value your time because you will realize that you can make endless amounts of money, but you can't produce more time. Make it your goal to send the shortest email possible when inviting successful people onto your show. Never say anything that is not true and if possible provide a link to something that shows that you actually know what you are talking about. Once they take the time to visit your website, if it is terrible, they are not going to say "Yes" to the invite. Make sure that your website is well-branded before inviting potential guests to visit your website. We shouldn't judge books by their covers, but we do.

Something you must do to really make the future guest's life easy is to send them over the questions you want to ask them BEFORE the interview so they can review them and edit them. Here is an example email you can send.

> Hey (Guest Name),
>
> I would like to interview you as a featured guest on (name of podcast)
>
> Throughought my career thus far I've been blessed to achieve X, Y, and Z. I would be honored to interview you. Please let me know if I can interview you.
>
> Sincerely,
>
> Clay Clark
>
> *Former U.S. Small Business Association Entrepreneur of the Year*

One of the worst things you can do when recording a guest is to not be prepared. You want to make sure to have sent over the questions well in advance of the interview so they can send over changes to the questions if needed. Also make sure to have their preferred method of recording. I use Skype (audio only).

Also, be prepared to carry the entire episode, just in case your guest is going to give one word answers. I've seen too many interviews end up as 5 minute episodes because the host had not done sufficient preparation for the episode and had nothing other to say than to ask the pre-prepared questions. Let the success of your episode be dependent on you (the things you can control) and not others, Ask questions, listen to their answers and respond to them. Remember this: Your podcast is an actual conversation.

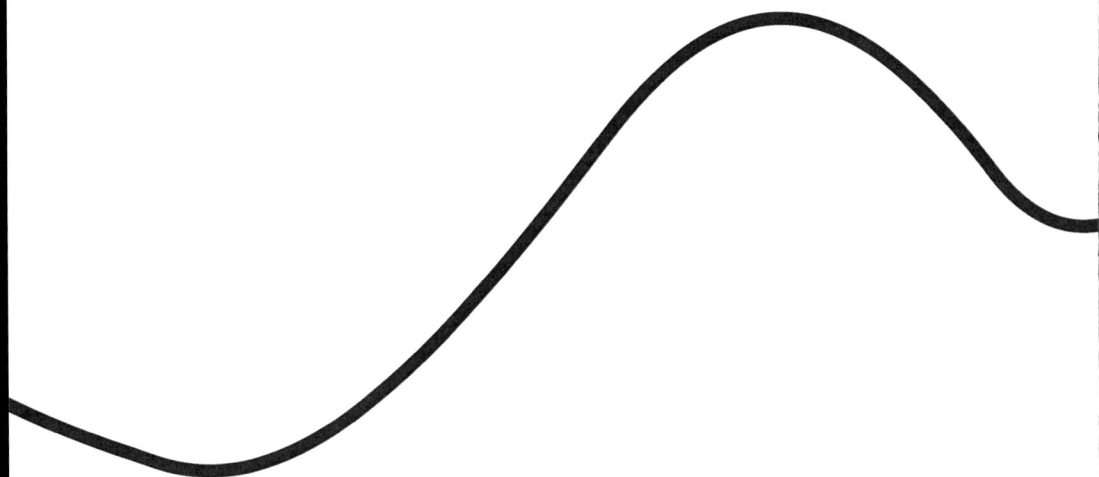

"WHEN MAKING SALES CALLS, INVITING GUESTS TO BE ON YOUR SHOW OR DOING ANYTHING THAT MIGHT INVOLVE YOU GETTING REJECTED YOU MUST DECIDE TO EITHER BE MENTALLY DUMB OR NUMB. REALLY DUMB PEOPLE ARE USUALLY OBLIVIOUS TO THE PAIN THAT REJECTION BRINGS AND A HEALTHY DOSE OF NOVOCAINE ALLOWS A DENTIST TO REMOVE A TOOTH FROM YOUR SKULL WITHOUT YOU CRYING. YOU MUST ACCEPT THAT YOU ARE GOING TO GET 300 REJECTIONS IN EXCHANGE FOR EVERY 'YES' YOU GET."

CLAY CLARK

(The man who has built several multi-million dollar businesses on a foundation of cold-calling.)

"SUCCESS IS A CHOICE."

NAPOLEON HILL

(The best-selling author if *Think and Grow Rich.*)

CHAPTER FOUR

WHAT GEAR IS NEEDED?

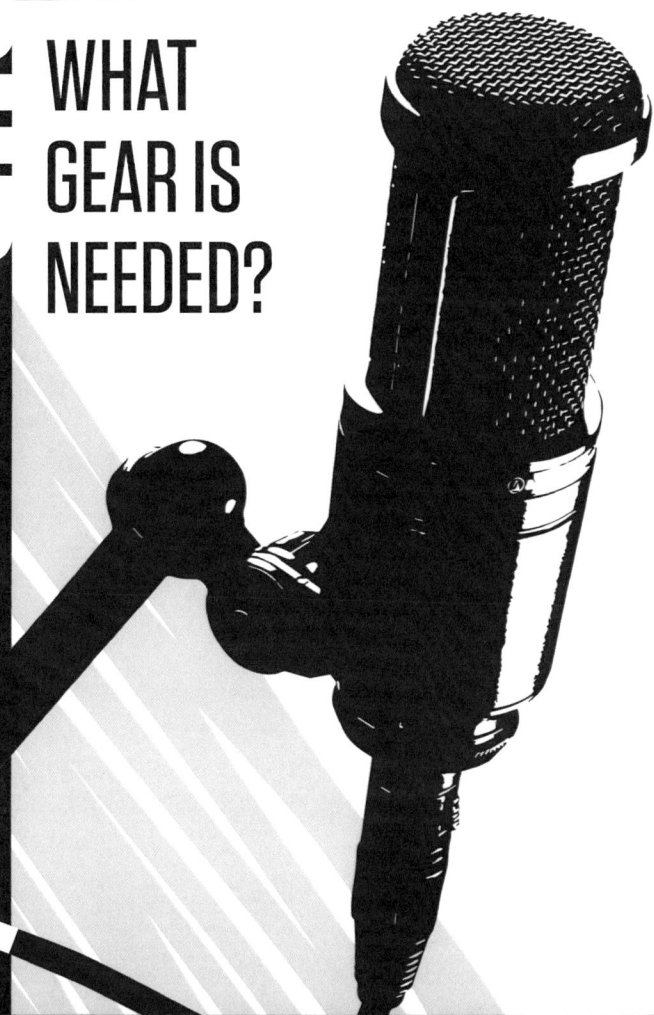

BUYING THE EQUIPMENT YOU NEED

Okay, you have decided to really record a podcast. You now know how to do it and how to invite quests, however, you are going to need the list of equipment that you actually need. While there are many different equipment setups out there, I recommend using the following equipment because I know it works and I have built a successful podcast platform using them.

"IF YOU'RE NOT A RISK TAKER, YOU SHOULD GET THE HELL OUT OF BUSINESS."

RAY KROC

(Founder of the McDonald's franchise.)

A COMPUTER

Because computers change and upgrade so often, I recommend buying a computer that was built in the last year or so that it is fast enough and updated enough to handle the software you are trying to run. Quit trying to install 2019 versions of Adobe onto your 1999 Microsoft DOS Computer.

Apple iMac

"BUY THE COMPUTER YOU NEED. DON'T MAKE EXCUSES, JUST BUY IT. WHEN ATTENDING COLLEGE WITH RYAN TEDDER (GRAMMY- WINNING ARTIST) I WAS ALWAYS AMAZED TO SEE THIS MAN ASSEMBLING A MUSIC STUDIO ITEM BY ITEM IN HIS DORM ROOM. WHILE THE OTHER GUYS ON THE FLOOR WERE SPENDING THEIR MONEY ON DATES, CARS, TVS AND VIDEO GAMES, THIS MAN WAS DELAYING GRATIFICATION AND CONCENTRATING ALL OF THE INCOME THAT HE WAS MAKING WAITING TABLES AT CHARLESTON'S TO BUY RECORDING EQUIPMENT. STOP WITH THE BS ABOUT NOT BEING ABLE TO AFFORD TO BUY THE COMPUTER YOU NEED AND MAKE THE TRADEOFFS YOU NEED TO GET IT DONE."

CLAY CLARK

(A man who worked at Target, Applebee's, and DirecTV simultaneously while living without air-conditioning during a hot Tulsa summer in order to be able to afford advertising in the Yellow Pages.)

I recommend Apple computers because they are powerful performers when it comes to audio recording. You also will want to use a computer that has multiple USB (Universal Serial Bus) ports for the microphone to plug into your computer.

A MIC

Starting out with a podcast, I recommend the Electro Voice RE320 Microphone that you can buy on Amazon.com.

> **"AS OF 2019, IF YOU ARE SERIOUS ABOUT PODCASTING, YOU SHOULD AT LEAST BELIEVE IN YOURSELF ENOUGH TO BUY A RE320 TO RECORD YOUR PODCAST. ALSO, ON A SIDENOTE, IF YOU ARE GOING TO BE SHOOTING WOMP RATS YOU REALLY NEED TO INVEST IN A QUALITY T-16 SKYHOPPER."**
>
> **CLAY CLARK**
> (The former U.S. SBA Entrepreneur of the Year and a man who loves Star Wars.)

I use the Electro Voice RE320 ($299) in the studio for myself and all of my in-person guests. You will need an XLR Cable for every microphone you have. For these microphones, you will need a Focusrite Scarlett 2i2 Interface for a permanent podcasting location ($149) or a Zoom H6 (which will require an SD card for recording.) for mobile podcasting ($349).

Focusrite Scarlett

If you are going to have more than 2 people at your permanent podcasting location podcast, get a Focusrite Scarlett 18i8 with 4 microphone inputs.

"BUY THE ELECTRO VOICE RE320 ($299) AND QUIT SAYING YOU CAN'T AFFORD IT. TURN OFF YOUR TV, QUIT GAMBLING, STOP SMOKING, STOP DRINKING, OR GET ANOTHER JOB... JUST GET IT DONE."

CLAY CLARK

(The man who was named Tulsa's Entrepreneur of the Year by the Chamber of Commerce at the age of 20 as a result of being able to live below his means and to go without TV, air-conditioning, and other things that most people choose not to go without.)

The Headphones - Headphones are SUPER important. It is vital that you are listening to yourself while recording as well as listening to your guests while they are talking. If your face is not touching the mic at all times, your levels and volumes will come out very poor and render the episode almost unusable. MAKE SURE YOU HAVE A QUALITY PAIR OF HEADPHONES. Which headphones should you get? Great question, I prefer the following:

Shure SRH440

» Shure SRH440 Studio Headphones – $99

» AKG K142 Headphones – $89

"DON'T SPEND $600 ON YOUR HEADPHONES. GET THE HEADPHONES WE RECOMMEND. SPEND YOUR WAR-CHEST ON PROMOTING, MARKETING AND ADVERTISING YOUR PODCAST."

CLAY CLARK

(The former U.S. Small Business Administration Entrepreneur of the Year.)

SOFTWARE

Just like the hardware, there are all kinds of programs you COULD use, but don't debate with me, and just use the following options.

PODCASTING SOFTWARE

- Adobe Audition – You can download the software at Adobe.com and it will cost you less than $20 per month to do so as of 2019.
- Skype – Computer-to-Computer Calling - Download the software at Skype.com. With Skype, you can make free computer-to-computer calls and affordable long distance calls which is the best solution for connecting with your guests for interviews.

"I TRULY BELIEVE THAT IN ORDER TO TRULY BE GREAT AT SOMETHING YOU HAVE TO GIVE INTO A CERTAIN AMOUNT OF MADNESS."

JOE ROGAN

(Host of one of the most downloaded podcasts of all time.)

"YOU'RE AN AVERAGE OF THE FIVE PEOPLE YOU ASSOCIATE WITH MOST, SO DO NOT UNDERESTIMATE THE EFFECTS OF YOUR PESSIMISTIC OR DISORGANIZED FRIENDS. IF SOMEONE ISN'T MAKING YOU STRONGER, THEY'RE MAKING YOU WEAKER."

TIM FERRISS

(Host of *The Tim Ferriss Show*.)

CHAPTER FIVE

ACTUALLY RECORDING YOUR AUDIO CONTENT.

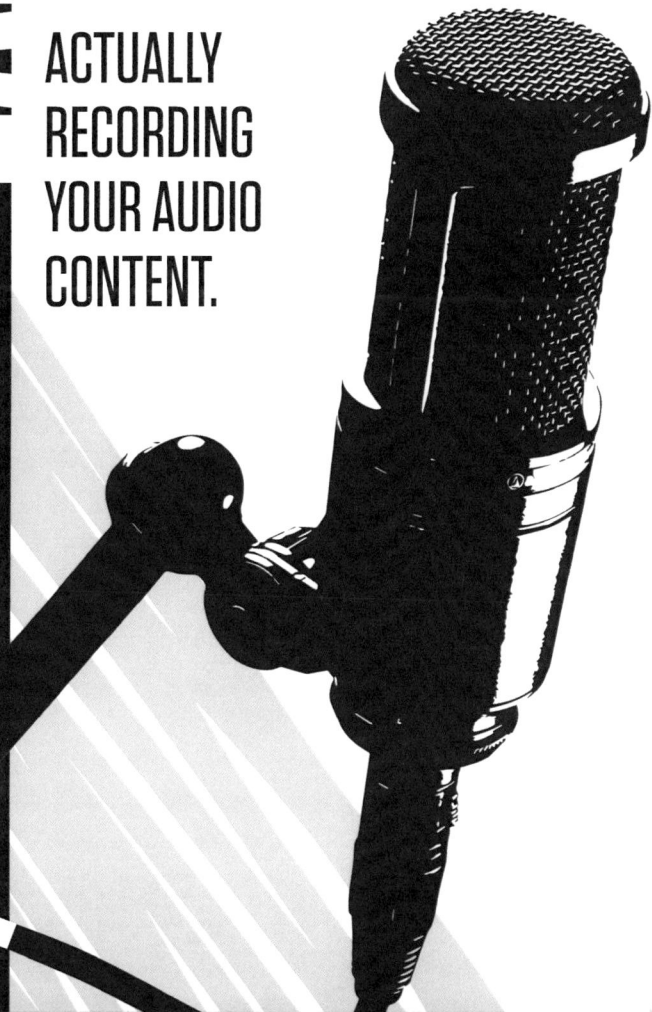

RECORDING YOUR PODCAST

"WE DON'T LIKE CHECKLISTS. THEY CAN BE PAINSTAKING. THEY'RE NOT MUCH FUN. BUT I DON'T THINK THE ISSUE HERE IS MERE LAZINESS. THERE'S SOMETHING DEEPER, MORE VISCERAL GOING ON WHEN PEOPLE WALK AWAY NOT ONLY FROM SAVING LIVES BUT FROM MAKING MONEY. IT SOMEHOW FEELS BENEATH US TO USE A CHECKLIST, AN EMBARRASSMENT. IT RUNS COUNTER TO DEEPLY HELD BELIEFS ABOUT HOW THE TRULY GREAT AMONG US—THOSE WE ASPIRE TO BE—HANDLE SITUATIONS OF HIGH STAKES AND COMPLEXITY. THE TRULY GREAT ARE DARING. THEY IMPROVISE. THEY DO NOT HAVE PROTOCOLS AND CHECKLISTS. MAYBE OUR IDEA OF HEROISM NEEDS UPDATING."

ATUL GAWANDE

(The Harvard Medical School professor and best-selling author of *The Checklist Manifesto: How to Get Things Right*.)

In order to improve the overall quality of your life by 2% easier I have taken the time to provide you with the detailed step-by-step instructions on how to set up your podcasting software. However, once you get the Adobe Audition program setup on your Apple Mac computer, you just need to invest at least 40 hours of your time using the program to become comfortable with it. 40 hours? Yes. Just like learning to ride a bike or learning to swim you must invest the time needed to master using the Adobe Audition software if you want to become a successful podcaster.

The checklist I have put together is working off of the assumption that you are using an Apple Mac computer. If you are not using an Apple Mac computer to record your podcast, you need to invest in the purchasing of an Apple computer today. If you are choosing to use a PC to record your podcast, this is a bad idea.

THE SPECIFIC STEPS THAT YOU MUST TAKE TO RECORD AN AUDIO PODCAST

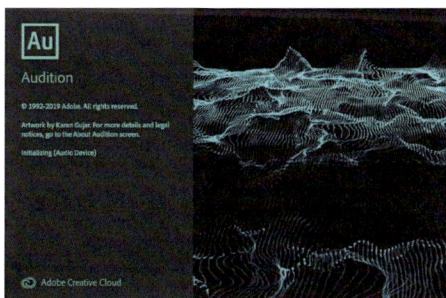

» **Open the Adobe Audition program**

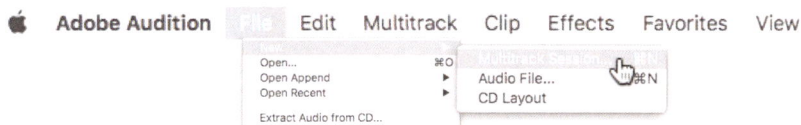

» **Go to File > New > Multitrack Session…**

» **Always name the "Session" with the accurate DATE and Podcast Company Name as shown below, then click "Browse…" Ex. "6.28.17 – Thrivetime Show"**

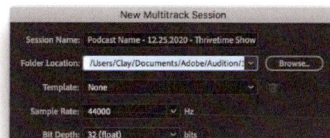

» **Click on "Documents" located on the left hand side. If it's your first time podcasting on this computer, create a folder by clicking "New Folder" button in the bottom left corner of the window. Name it the podcast name. If you have previously podcasted on this computer, select the podcast's folder in "Documents."**

» **Set the setting as indicated in the screenshot below and click "OK."**

» **Click "Adobe Audition CC" in top Navigation Bar > "Preferences"... and Click "Audio Hardware."**

» **Verify that your microphone (or Scarlett 2i2 USB) is selected for Default Input. For Scarlett 2i2 it should be set to both default Input and Output, and then click "OK."**

» **For the Blue Yeti Microphone setups: On Track 1, make sure the right arrow is set to Mono > Blue Yeti Microphone. Only microphones with headphone jacks could be set to both input and output.**

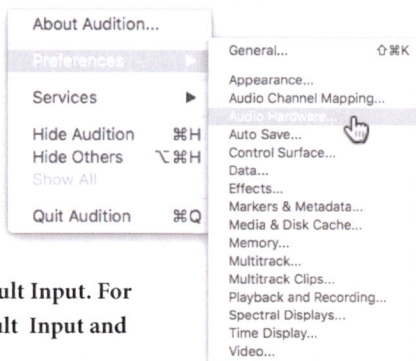

FOR MULTIPLE MICROPHONE SETUPS WITH THE SCARLETT:

1. On Track 1, make sure the right arrow is set to Mono > [01M] Scarlett 2i2 USB: Input 1. On Track 2, make sure the right arrow is set to Mono > [02M] Scarlett 2i2 USB: Input 2.

2. Click the R so it turns red for Track 1, and click the R so it turns red for Track 2 (if there are multiple microphones).

3. Press "Red Circle" at bottom of window to start recording an episode:

4. Record for 10 minutes:

5. State which Podcast Episode # it is and Podcast/Company Name

6. State the topic within the first two sentences

7. Introduce Yourself and your business.

8. Make sure to say the search engine optimization keyword that you are

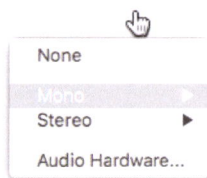

focused on at least 6 times throughout each 10 minutes of your podcast.

9. Once 10 minutes of recording is complete, click the square icon at the bottom of the window. **STOP**

10. Once stopped, if you are going to record another episode, move the cursor (blue upside down triangle) forward by clicking and dragging it.

11. To export the episode, start by zooming out so you can see an entire episode. You can do this by pressing the minus key on the keyboard (to the right of the 0).

12. Drag the Blue Playhead (upside down triangle) by clicking and dragging to the beginning of where an episode starts.

13. Press the letter i:

14. Drag the Blue Playhead (upside down triangle) by clicking and dragging to the beginning of where an episode ends.

15. Press the letter o:

16. With the episode highlighted, select File > Export… > Multitrack Mixdown > Time Selection…

17. Click Browse…

18. Name the File under "Save As"… as follows:

19. Podcast Number - Podcast Name

20. Example: 16 - ThrivetimeShow

21. Click "Documents"

22. Click your "Podcast Exports" folder

23. Click "Save"

24. Make sure the settings are the same as below

25. Click "OK"

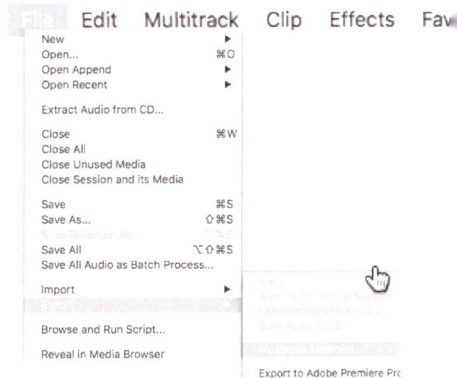

"PEOPLE CHOOSE UNHAPPINESS OVER CERTAINTY."

TIM FERRISS
(Host of *The Tim Ferriss Show.*)

CHAPTER SIX

XIS

HOSTING
YOUR
PODCAST.

Once you have recorded your podcasts, you are going to want to put them somewhere where humans can actually listen to and consume them, am I correct? There are tons of different places that you COULD use to host your podcast but don't overthink it. I recommend having a branded website to post your podcasts to and hosting your podcast on Libsyn.com

Do not get stuck or overwhelmed. If you can't figure out how to set up your own website or how to host your content on Libsyn, attend a ThrivetimeShow.com workshop so that our team can show you inperson how to setup your podcast. It's not your fault if you struggle to understand a new concept. However, it is your fault if you decide to not allow yourself to attend our *Thrivetime Show* business conferences (get your tickets at Thrivetimeshow.com/business-conferences).

CREATING YOUR WEBSITE

Once you create a website, it is easy to use Libsyn to add podcasts to your website, but in order to create a website, I recommend using WordPress. Why? Google loves it, it is the most easily updated website platform out there and super user friendly, so I love it too.

WORDPRESS WEBSITE – FREE –WORDPRESS.ORG

Bonus Note: If you need someone to build out a WordPress site for you, optimize it for Google, AND set it up for podcasts, we can do it for you for less than what it would cost for you to employ one $8.25 hour employee. To learn more about having our team help you to set up your website, e-mail us today at info@thrivetimeshow.com.

SMART PODCAST PLAYER PLUGIN
SMARTPODCASTPLAYER.COM

This is the podcast player that many top podcasters in the world use including John Lee Dumas (host of the *EOFire Podcast*), and many others. Don't over think this and decide today what podcast player you will use.

LIBSYN.COM

This is where you will host your podcast and it will distribute your shows to the top destinations on the planet including iHeartRadio, Pandora, Spotify, Radio.com, Stitcher, and many others. When signing up for Libsyn.com there are 2 plans to choose from but to start at the lower package at $5 a month will be just fine.

Libsyn is also where you will go to create an RSS feed which is what you will need in order to have your podcast featured Apple podcasts.

CHECK OUT THE SUPER LIBSYN
UPLOADING CHECKLIST:

» Add Media File to iTunes

» Right Click on File and Click Get Info

» Click Artwork

» Click Add Artwork

» Add Appropriate Artwork

» In Libsyn

1. Hover over the Content tab

2. Click Add New Episode

3. Click Add Media File

4. Click Select from FTP/Unreleased or Upload your file

5. Click the Select button to the right of the file name for the episode

> a. You may preview the episode by clicking on the preview button to the right of the file name

15-KFAQ-Thrive-Time-Show-09-8-16-Entire-Show-with-Ads.mp3   2017-12-07 22:12:50   **⟲ PREVIEW**   **✓ SELECT**

6. Click Go To Details

7. Copy/Paste the Title from the existing episode on your website

8. Copy/Paste the Description (podcast excerpt) from the existing episode on your website

9. Type name of the podcast in the Category field

10. Enter keywords

11. Permalink Points to - Select Custom URL

> a. Paste the link to the page on the website

12. Copy/Paste the Title into the iTunes Title

13. Copy/Paste the Description into the iTunes Description

14. Enter Season #

15. Enter the Episode number

16. Choose Clean for the Rating

17. Enter iTunes Author (Author of the Podcast)

18. Click Go To Artwork

19. Paste the artwork link

20. Click Go To Scheduling

21. Click the Basic Release/Expiration tab

22. Click Set new release date

23. Enter the podcast date it was published on

24. Check the Update ID3 tags box

25. Click Publish

26. Copy the Direct Download URL

27. On your website, paste the direct download URL in the Podcast URL field

28. Under the Podcast Episode section, check Modify existing podcast episode

29. Paste the direct download URL in the Media URL field

30. Click Verify URL

Only applies when using BluBerry

31. Verify that the podcast is in a category

32. Click Update

"YOU SHOULDN'T FOCUS ON WHY YOU CAN'T DO SOMETHING, WHICH IS WHAT MOST PEOPLE DO. YOU SHOULD FOCUS ON WHY PERHAPS YOU CAN, AND BE ONE OF THE EXCEPTIONS."

STEVE CASE

(Co-founder of AOL.)

"A PERSON WHO NEVER MADE A MISTAKE NEVER TRIED ANYTHING NEW."

ALBERT EINSTEIN

(A legendary physicist who pushed President Roosevelt to allow him to create the atomic bomb before the German's completed their atomic weapon. Without Albert Einstein, America would have lost World War II.)

"THE MOST PRECIOUS THING WE ALL HAVE WITH US IS TIME."

STEVE JOBS
(The Co-founder of Apple, the founder of NeXT, and the former CEO of PIXAR who saved the company.)

CHAPTER SEVEN

SUPER MOVES FOR LAUNCHING YOUR PODCAST.

Once you have invested copious amounts of your time to learn how to properly and technically record your podcast it is now time to get your podcast out to the world and in front of your ideal and likely buyers so that they can listen to it, love it and hopefully share it with their friends, family and their network. However, before we begin discussing the marketing of your podcast, it is SUPER IMPORTANT for you to embrace the mindset that, nobody is going to wake up with a burning desire to pay you or to find your podcast. It is up to YOU and YOU ALONE to get your message and your audio "dojo of mojo" out to the masses.

"BUILD IT, AND THEY WILL COME" ONLY WORKS IN THE MOVIES. SOCIAL MEDIA IS A "BUILD IT, NURTURE IT, ENGAGE THEM AND THEY MAY COME AND STAY."

SETH GODIN

(A *Thrivetime Show* Podcast guest, *New York Times* best-selling author and the man who sold his company Yoyodyne to Yahoo! for a reported $30 million.)

It is truly incredible how many people now listen to podcasts on a daily basis. In fact according to a recent article published in the *New York Times* by Jaclyn Peiser, now one out of every three people listen to a podcast every month. That is HUGE! Podcast listeners today listen to podcasts via iTunes, Spotify, iHeartRadio, Podbean, YouYube, and a variety of other players. Thus, in order to get in front of your ideal and likely buyers it is VITALLY IMPORTANT that you work to insure that your podcast shows up on as many of the player platforms as possible.

FUN FACTS

FUN FACT

"The number of podcast listeners has increased sharply this year, according to a new report. More than half the people in the United States have listened to one, and nearly one out of three people listen to at least one podcast every month. Last year, it was more like one in four. .."What moved the needle is Spotify adding podcasts," Mr. Webster said of the music streaming service, which increased its podcast offerings last year and recently acquired Gimlet Media, the studio behind the popular podcasts "Crimetown" and "Reply All," and Anchor, which makes tools for recording and distributing podcasts."
- *Podcast Growth Is Popping in the U.S., Survey Shows* - Jaclyn Peiser
- https://www.nytimes.com/2019/03/06/business/media/podcast-growth.html

Don't get mentally hung up on becoming the king of iTunes or the Sultan of Spotify. You must focus on creating great, engaging and compelling content that listeners will love while making sure that on a technical basis that your podcast checks all of the boxes so that users can actually find your podcast when searching using their favorite podcast player.

> **"I'M CONVINCED THAT ABOUT HALF OF WHAT SEPARATES THE SUCCESSFUL ENTREPRENEURS FROM THE NON-SUCCESSFUL ONES IS PURE PERSEVERANCE."**
>
> **STEVE JOBS**
> (The Co-founder of Apple, former CEO of Pixar, and the founder of NeXT.)

"IN A CROWDED MARKETPLACE, FITTING IN IS FAILING. IN A BUSY MARKETPLACE, NOT STANDING OUT IS THE SAME AS BEING INVISIBLE. REMARKABLE PRODUCTS AND PEOPLE GET TALKED ABOUT. BORING IS INVISIBLE."

SETH GODIN

(Bestselling author of *Purple Cow*. He used $20,000 in savings to found Seth Godin Productions, primarily a book packaging business, out of a studio apartment in *New York City*. He then met Mark Hurst and founded Yoyodyne. In 1998, he sold Yoyodyne to Yahoo! for about $30 million and became Yahoo's vice president of direct marketing.)

"YOU TRY TO DO THE BEST WITH WHAT YOU'VE GOT AND IGNORE EVERYTHING ELSE. THAT'S WHY HORSES GET BLINDERS IN HORSE RACING: YOU LOOK AT THE HORSE NEXT TO YOU, AND YOU LOSE A STEP."

JIMMY IOVINE

(Legendary record producer who has sold over 250 million albums and the genius mastermind both Interscope Records and Beats Headphones.)

Each year, I personally speak with hundreds of people who tell me that I have inspired them to create their own podcast, however sadly whenever I take the time to actually follow-up I almost always hear the following for some reason or another:

» "Yeah, I actually didn't have time…"

» "Well, I decided not to launch that podcast…"

» "With the baby and all this hasn't been a great time for me to start…"

» "I just got overwhelmed and we stopped recording it…"

» "I just ran out of time…"

» "I want to do it, but with all of the demands of our build out I just couldn't find the time…"

I believe in you and I know that you are capable of having both the mental capacity and the tenacity needed to launch your podcast, but you must schedule the time into your calendar to record, upload and optimize your content or you will NEVER do it.

SO I AM ASKING YOU NOW…

When are you going to write your show outlines? _____
_____

When are you going to record your content? _____
_____

When are you going to upload your content? _____
_____

When are you going to optimize your content? _____
_____

When are you going to reach out to potential guests? _____
_____

It is of EPIC IMPORTANCE that you actually block off the time in your calendar for these activities or you simply will not EVER get it done and you will just run around feeling overwhelmed with show ideas that never get turned into reality.

CREATE QUALITY EPISODES

This probably goes without saying, but you are going to have to actually produce content that people want to hear. How? Great question!

> »   Hook Your Subscribers

In each episode, you are going to want to ask your listeners to do two things: Go to iTunes to subscribe and leave a review

Refer back to a previous episode, or get them excited for the next one by sharing about an upcoming topic or guest.

> »   Maintain a High Download Count

> »   Release With Multiple Episodes

I recommend getting at least 10 episodes completed and posted before releasing, and 10 more (20 total) ready to be posted regularly after you launch on iTunes. This gives your new listeners an opportunity to listen to more episodes after they fall in love with you (and I know they will), and they will be more likely to become a subscriber. It's also a great way to drive more downloads while separating yourself from all of the other organizations and hosts (Apple has over 500,000 podcasts submitted to them) that have started a podcast because they thought it would be fun and easy but quit just a few weeks into it.

If you have 10 episodes already posted and you get 10 new subscribers, that can amount to 50 new downloads! However, if you start with only 1 or 2 episodes, and have 10 new subscribers, that would result in only 10-20 downloads.

ITUNES IS IN CONTROL - FOLLOW THEIR RULES

If you are not submitting a quality photo, a complete description, good episode titles and description, then iTunes is likely to keep you out of the New & Noteworthy section because they do not want to promote a poor-looking album image. Make sure to download their list of requirements and follow it!

> » **Get Reviews**

Once you submit to iTunes, it will be important for you to personally ask your listeners for reviews on your podcast episodes. The more engagement that your podcast has on iTunes, the more likely your podcast will be featured on the New & Noteworthy section. Make sure you always ask for objective reviews during your shows and don't pay people to review you highly.

"DO NOT BE EMBARRASSED BY YOUR FAILURES, LEARN FROM THEM AND START AGAIN."

RICHARD BRANSON

(The dyslexic founder of Virgin Group who started his first business, a newspaper, after dropping out of high school in spite of his dyslexia.)

With all of these factors contributing to your overall ranking on iTunes, the last point and most important thing to remember is consistency. Google will not promote your business to the top of search results if you are spamming them, and iTunes will not promote you if you are caught spamming them either. If you get a bunch of reviews immediately, but then never ask or receive another ever again, then you will be less likely to retain a high ranking. Make an effort to ask for real reviews from real listeners consistently and your efforts will pay off.

"A SURE-FIRE WAY TO PREDICT THE FUTURE IS TO TAKE NO ACTION AT ALL. WHEN YOU DO NOTHING, YOU GET NOTHING."

PAT FLYNN

(The founder of the *Smart Passive* Income Podcast.)

"SELF-MASTERY IS A DIFFERENT STEP, BUT A NECESSARY ONE. EO FIRE CAME TO LIFE BECAUSE I SET AND ACCOMPLISHED A GOAL, AND BUILT A SEVEN-FIGURE-A-YEAR BUSINESS BECAUSE I MASTERED PRODUCTIVITY, DISCIPLINE, AND FOCUS."

JOHN LEE DUMAS
(The founder and host of the *Entrepreneur On Fire* podcast.)

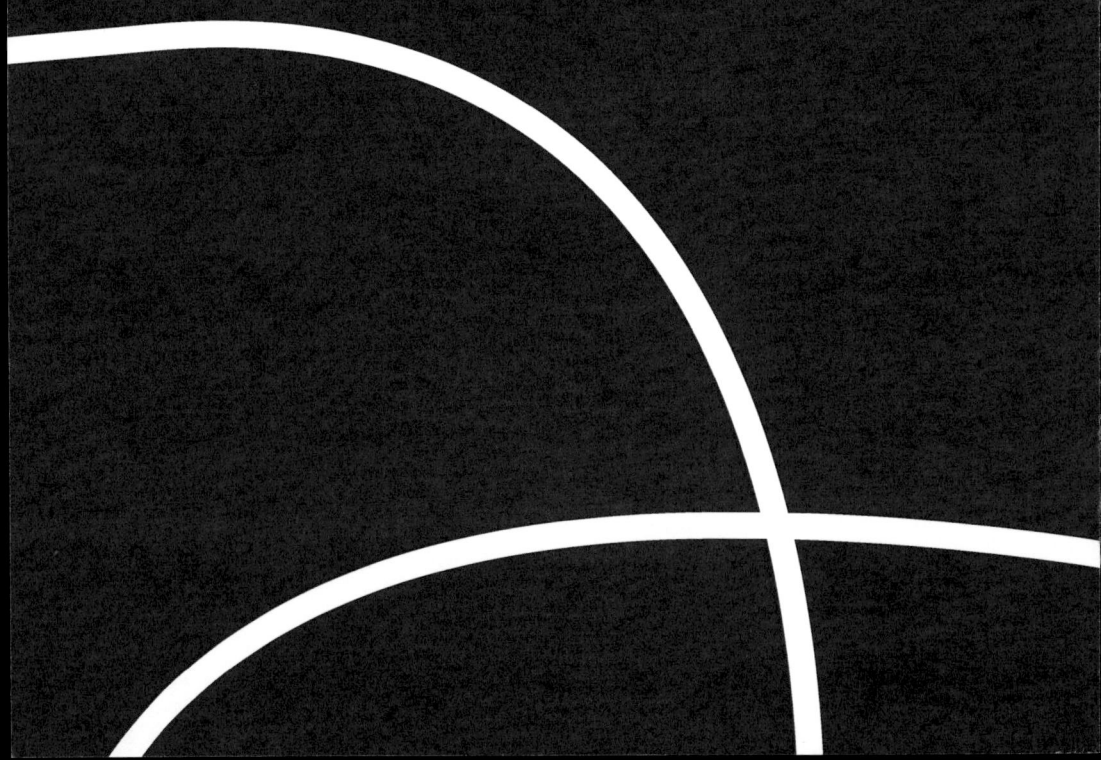

CHAPTER EIGHT

SUBMITTING YOUR PODCAST TO ITUNES.

It's time: time to launch on iTunes. There are not a lot of things that need to be said about this, so here is the simple checklist for how to get your podcast onto iTunes.

"HALF THE BATTLE IS SELLING MUSIC, NOT SINGING IT. IT'S THE IMAGE, NOT WHAT YOU SING."

ROD STEWART

(Born and raised in London, he is of Scottish and English ancestry. Stewart is one of the best-selling music artists of all time, having sold over 100 million records worldwide. He has had six consecutive number one albums in the UK and his tally of 62 UK hit singles includes 31 that reached the top ten, six of which gained the #1 position. Stewart has had 16 top ten singles in the US, with four reaching #1 on the Billboard Hot 100.)

CREATE AN APPLE ID

The first thing you will need to distribute your podcast on iTunes is to create an Apple ID if you do not have one. Most people will already have a personal Apple ID from using iTunes, the App Store, or buying an Apple Product but if you do not, go through the process of setting this up.

LOG INTO PODCASTS CONNECT - PODCASTSCONNECT.APPLE.COM

Login to Podcastsconnect.apple.com with your existing Apple ID and follow the steps to submit your podcast with your RSS feed (that you set up in Libsyn.com to Apple Podcast. Be sure to have the following filled out

» Podcast Name

» Podcast Album Artwork (needs to be 2000px by 2000px)

» Podcast Subtitle

» Podcast Summary (what your show is about and why people should listen)

» Your summary cannot exceed 4,000 characters in length

» iTunes Category

This is where you will need to decide where your podcast fits. Make sure to select a main category (for example - Business) and also a subcategory (example - Management & Marketing)

Decide if your podcast needs to be marked "explicit" (will there be swearing?) or not.

» Podcast host or author name

» Email for the podcast

Submit your Podcast RSS Feed - Once the feed information has been filled out, you can submit your podcast to iTunes at podcastsconnect.apple.com. Once it is submitted, it will take a day or two to get approved and launched into the iTunes Store.

BONUS NOTE: Once getting your podcast approved and on iTunes, you can submit your podcast for consideration to the Apple team for the "Featured Podcast" section. You can submit your podcast to be considered by the Apple team today by visiting - https://itunespartner. apple.com/contact/?appleId=&content_type=podcasts&content_type_ name=&topic=public-promotion-request

"I THINK IT IS POSSIBLE FOR ORDINARY PEOPLE TO CHOOSE TO BE EXTRAORDINARY."

ELON MUSK
(The man behind SpaceX, Tesla, SolarCity, Paypal, Etc...)

"EXCELLENCE IN ANYTHING INCREASES YOUR POTENTIAL IN EVERYTHING."

JOE ROGAN

(Host of one of the most downloaded podcasts of all time.)

"WHETHER YOU THINK YOU CAN OR YOU THINK YOU CAN'T YOU ARE RIGHT."

HENRY FORD
(The man who founded Ford Automotive.)

CHAPTER NINE

POWER
PRINCIPLES FOR
ESTABLISHING
AND GROWING
YOUR AUDIENCE.

Now that you are actually recording your content, this is a GREAT THING. You are a doer and not just another intender. BOOM. Now, it's important that YOU focus on growing your audience. However, it's important that you are very self-aware. If your podcast is terrible, focus on making it not terrible. Then focus on making it good. Then focus on making it great. Then focus on marketing your great content. Do not go out there and spend thousands of dollars marketing something that is terrible.

"MARKETING IS A CONTEST FOR PEOPLE'S ATTENTION."

SETH GODIN

(Author of numerous *New York Times* best-selling books.)

So let's work off the assumption that your show is absolutely terrible right now. Why? Because I was terrible when I started recording my show and most people are terrible when they first start. So how do you improve?

Schedule a time every day to listen to your own show and podcasts to verify they are not terrible and to ask yourself the following questions:

» How you could improve your show?

» Are the introductions to your shows good?

» Do you end your shows well?

» Are your transitions good?

»   Are your sound effects good?

»   Do your guests connect with your audience?

»   Do you ask questions well?

»   Do you talk too fast?

»   Can people actually hear the words that are coming out of your mouth from a technical perspective?

Once you decide that your content is good enough to promote via social media I recommend being consistent. Again, consistency in creating a valuable social media account where you post every time an episode that is live and in a timely fashion is critical to your success as well as having a thumbnail that is interesting and that makes people want to click.

Make sure you are posting to your personal social media accounts as well as the brand's social media account. Some of the clients that I have worked with have felt shameful not wanting to push their episode to their friends on social media for fear of them being "unfriended." Hilarious! To that I would say that you can expect zero downloads and subscriptions. I would not recommend spamming your contacts with content that isn't valuable, but when you start your business, brand, or podcast, you ARE the brand. Utilize the tools and the network of people that you have at your disposal to promote your podcast. One tool you can use (if you are using WordPress) to post your content to your social media channels automatically is called "Zapier"

EMAIL WHEN A NEW EPISODE IS LIVE

Another thing you can do to engage your audience and increase the number of downloads of your podcast is to email a subscriber list. You can create this

list by adding a form as a popup on a website, or a form that a listener fills out. One of the ways to get people to want to fill it out the forms on your website is to give something away for free. Many successful podcasters have created an email list that is interested in knowing when the next episode is live so that they can be sure to listen to it. You can ask people to opt-in to an email list that notifies your tribe when a new episode is live. This is a way to quickly grow an audience and ensure that your listeners are getting the newest content that is released (and increasing your downloads). NOTE: We do not recommend sending an email every single day if you are releasing podcasts daily. Weekly or monthly is the most amount of e-mails that I would recommend to send your ideal and likely buyers and subscribers

EMAIL THE GUEST WHEN A NEW EPISODE IS LIVE

One of the most valuable things that you can do when you release a new episode is to engage any guest that is on the episode and to leverage their social media following. You will find that these individuals are some of your biggest advocates and best marketers.

EVERYONE'S FAVORITE TOPIC? THEMSELVES!

In the shameless attempt to promote themselves, your guests will share their interviews with their audience and followers. This further provides valuable content to their fans and validates them as an expert among their own followers as well while opening up new potential listeners that have not heard of you before they find out about your great podcast.

YOUR GUEST'S NETWORK

By asking your guest to promote your podcast episode to his or her audience, you are going to reach an entirely new group of people that will hear your podcast, and that get to experience your wisdom for the first time. With the power of a referral from a guest, you will arrive in front of many new ears hearing your podcast for the first time and when you are first starting out, there is a great probability that your guest's audience is already much larger than yours so getting out in front of them is an easy way to help your podcast expand its reach.

> "DON'T EMAIL YOUR GUESTS IN A SPAM LIKE, CLICK-FUNNEL BS KIND OF WAY. TREAT YOUR GUESTS LIKE SOMEONE THAT YOU ARE DATING AND ACTIVELY TRYING TO WOO. BE THOUGHTFUL AND INTENTIONAL."
>
> **CLAY CLARK**
> (The former U.S. Small Business Administration Entrepreneur of the Year who started his first business in his parent's basement and then he scaled it out of his college dorm room, DJConnection.com.)

When you reach out to a guest to let them know that his or her podcast is live, this is what you need to remember and include: Email them early in the morning of the day that the episode is live, text them early in the morning of the day that the episode is live if you have their number. Do whatever you need to do in order to make sure that they see and know that their podcast is live. You should also tag them on social media when you post the podcast.

When you email them or text them, make sure you include a link to the podcast as well as a description of the show that they can use. You want to make sharing the podcast as easy on them as possible.

ENGAGING ON SOCIAL MEDIA

Lastly, you will want to engage with your audience on social media. Between episodes, you will have listeners that want to connect with you further. Provide them with a platform where they can ask questions, connect with other listeners, and get additional content. Whether this is your website (like ThrivetimeShow. com), or another social media platform, make sure that whatever platform you direct them to is being monitored and actually answered.

Often I will see podcast hosts drive traffic to a social media platform and then they either not respond at all to people's questions or they have chosen to automate their response.

"THE MEDIA WANTS OVERNIGHT SUCCESSES (SO THEY HAVE SOMEONE TO TEAR DOWN). IGNORE THEM. IGNORE THE EARLY ADOPTER CRITICS WHO NEVER HAVE ENOUGH TO PLAY WITH. IGNORE YOUR INVESTORS WHO WANT PROVEN TACTICS AND PREDICTABLE INSTANT RESULTS. LISTEN INSTEAD TO YOUR REAL CUSTOMERS, TO YOUR VISION, AND MAKE SOMETHING FOR THE LONG HAUL. BECAUSE THAT'S HOW LONG IT'S GOING TO TAKE."

SETH GODIN

(The man who continues to be the legendary marketing expert and best-selling author of numerous books.)

Now do not misunderstand me here, you should not spend endless amounts of time on social media. Proactively schedule the time every day that you will respond to messages, answer questions, and engage with your audience. Don't let this take copious amounts of your time. You should do all of your critical thinking at one time and schedule your posts using Hootsuite.com.

As part of *The Thrivetime Show* Podcast, I knew that we needed daily content for our platforms on social media in addition to the daily radio shows that we broadcast.

I used Hootsuite.com to schedule the posts to all of our social media platforms and so I did not think about what I was posting for the day ever again. In addition, I can still post timely and relevant content as it comes up, but now I don't need to worry about not posting for a day.

"THE WAY TO GET STARTED IS TO QUIT TALKING AND BEGIN DOING."

WALT DISNEY

(The man who started the Disney empire after famously losing it all twice. Imagine how he must have felt, but he did not quit.)

"CONSISTENCY IS SUPER IMPORTANT WITH ANY TYPE OF MARKETING. WE DESIGNED OUR SOCIAL MEDIA POSTS TO BE CONSISTENT AND RELEVANT TO OUR NICHE FOCUS OF IDEAL AND LIKELY BUYERS (ENTREPRENEURS) AND THEN WHEN SOMETHING UNPREDICTABLE AND RELEVANT HAPPENS WE WILL MAKE A SOCIAL MEDIA POST ABOUT IT (THE NFL'S PATRIOTS WINNING A GAME, ONE OF OUR BUSINESS COACH CLIENTS EXPERIENCING A BIG WIN, OPENING UP ANOTHER ELEPHANT IN THE ROOM MEN'S GROOMING LOUNGE LOCATION (EITRLOUNGE. COM) OR OXIFRESH.COM CELEBRATING THE 100,000TH GOOGLE REVIEW.)"

CLAY CLARK

(The former U.S. Small Business Administration Entrepreneur of the Year for the State of Oklahoma.)

"PEOPLE WHO KNOW WHAT THEY'RE TALKING ABOUT DON'T NEED POWERPOINT."

STEVE JOBS

(The Co-Founder of Apple, the founder NeXT, and the former CEO of PIXAR.)

"IF YOU DON'T LOVE IT, YOU'RE GOING TO FAIL."

STEVE JOBS

(The Co-founder of Apple, former CEO of PIXAR, and the founder of NeXT.)

CHAPTER TEN

HOW TO MONETIZE YOUR PODCAST.

A few years back, I was helping a client of mine who was having a rough time understanding how to turn her passion into a profitable business. As a holistic health coach in her mid-twenties, she sincerely wanted to help people by providing them with information on what they can do to optimize their life. She wanted to do this full-time, but was unsure how this could actually produce enough income to replace what she was currently doing.

As we were going through what her goals were, we decided that podcasting might be a good vehicle to get her where she wanted to go. She loved this idea and lit up with enthusiasm. As an avid podcast listener herself, She LOVED podcasts but just starting a podcast does NOT mean that you can immediately start making money from it. For her, it was important for her to still focus on creating some kind of income while she grew her podcast audience size.

I have to warn you, podcasting is not a "get rich quick" scheme (in fact, there is no such thing, but we can argue about that another time in another book). However, you CAN make money from it if you are diligent. Everyone who starts a podcast has the dream of making money from it but few people actually know how to grow the podcast to the size needed to actually produce revenue as a result of their podcast.

> "PROFIT IN BUSINESS COMES FROM REPEAT CUSTOMERS, CUSTOMERS THAT BOAST ABOUT YOUR PROJECT OR SERVICE AND THAT BRING FRIENDS WITH THEM."
>
> **W. EDWARDS DEMING**
> (Renowned economist and best-selling author)

GET SPONSORSHIPS

The most popular way to create revenue from your podcast is having a sponsor for the show who will agree to pay you money in return for their name being mentioned. Typically companies will approach you about sponsorship opportunities once you hit a large enough audience size. Make sure that when you accept a sponsor, you actually agree with their values, morals, and the products and services that they provide.

You can also attempt to go out and find your own sponsors and you can do that by:

1. Deciding what advertiser would be a good fit for your show (just make a big ol' nasty list of them)

2. Actually contact them!

Start by sending your potential advertisers an email. In this email you should include:

- Your name and an overview about your podcast

- A link to a recent podcast episode

- Your audience size and the demographic makeup of your listeners

- For your first couple of advertisers, I recommend making it a NO BRAINER (cheap) for them because no one wants to be the first to advertise on a show.

After you email them, if you have not heard back, try actually calling them and talking to them on the phone. Once you talk to them, in order to truly stand out, send a handwritten letter thanking them for their time.

When working with sponsors, there are 3 common "spots" you can sell on your podcast and those include: pre-roll, mid-roll, and outro advertisements.

THE PRE-ROLL

The pre-roll is exactly what it sounds like which is an ad that runs "pre" or before the content of the podcast itself. This is typically about 15 seconds long and after you introduce the topic of the podcast.

MID-ROLL

The mid-roll ad is anywhere from 50%-70% of the way the podcast and is an extended advertisement (typically 30-45 seconds long). You would charge more for this as this is the most prominent advertisement in the podcast episode.

OUTRO

This is at the very end of the episode and is typically the same length as the pre-roll (15 seconds long). You can end it by saying "this episode has been brought to you by…."

HOW MUCH MONEY CAN YOU CHARGE?

At the end of the day it is completely up to you what you charge but the "standard" in the world of pocasting is about $20 per 1,000 downloads and outros and $25-30 per 1,000 people that listen for the midroll advertisment.

For the people that had to take algebra 3 times like myself, I have done the math for you:

> » $20 per 1,000 downloads x 27,000 Downloads = $540 per show

You can see how fast those numbers add up and the benefit that producing more shows could be to you and your wallet.

"I TRULY BELIEVE THAT IN ORDER TO BE GREAT AT SOMETHING YOU HAVE TO GIVE IN TO A CERTAIN AMOUNT OF MADNESS"

JOE ROGAN

(One of the most listened to and downloaded podcasters of all-time.)

"REMEMBERING THAT I'LL BE DEAD SOON IS THE MOST IMPORTANT TOOL I'VE EVER ENCOUNTERED TO HELP ME MAKE THE BIG CHOICES IN LIFE."

STEVE JOBS

(The Co-founder of Apple, former CEO of Pixar, and the founder of NeXT.)

CHAPTER ELEVEN

IT IS FINISHED. YOU'VE DONE IT.

Congrats, you are now on your path to becoming a bonafide podcaster. I sincerely know that you can apply these steps in your own podcast to achieve the success that you are wanting. However, success is not created by luck. There is no such thing as "get rich quick." You are the master of your own fate. Podcasting is a world that I have gotten very excited about in the past 5 years because of both the time and financial freedom it can provide. With a podcast, you are no longer shackled to the mere 1-to-1 ratio of people that you can help. Now, you can scale your vision and provide practical training to millions across the world. If you have any questions about this book or would like to connect, please email us at info@ThrivetimeShow.com. This is your year to THRIVE. I'm excited for your success.

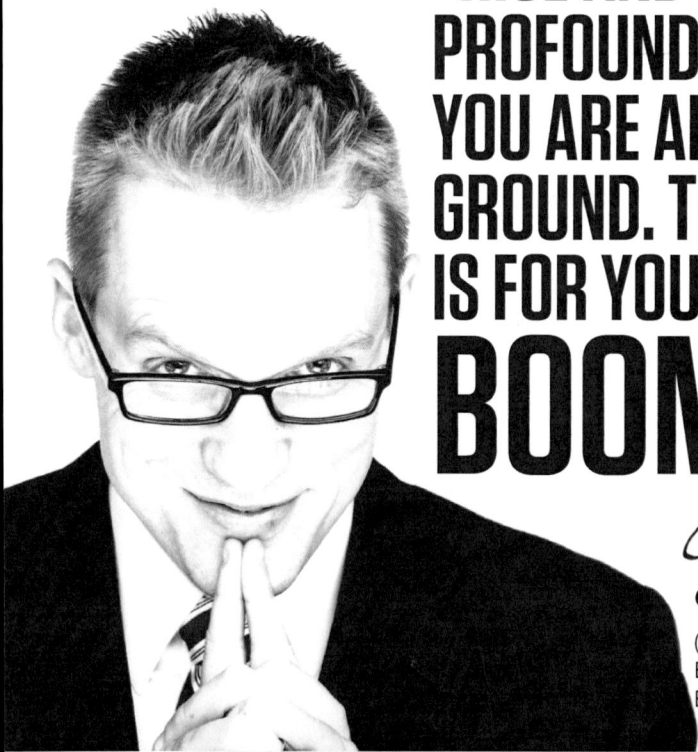

"RISE AND GRIND! IT'S PROFOUND EVERYDAY YOU ARE ABOVE THE GROUND. THIS BOOM IS FOR YOU. BOOM!"

CLAY CLARK

(Former U.S. Small Business Administration Entrepreneur of the Year.)

ACTION ITEMS

»   Pass on what you've learned by writing a Google Review. Type in "Thrivetime Show Jenks" and write that review today!

»   Don't miss a radio show or podcast. Subscribe and review us on iTunes at ThrivetimeShow.com.

»   Get all of the interactive downloadables by signing up today at ThrivetimeShow.com.

WANT TO KNOW EVEN MORE?

CHECK OUT ALL OF CLAY'S BOOKS

START HERE
The World's Best Business Growth & Consulting Book: Business Growth Strategies from the World's Best Business Coach.

DON'T LET YOUR EMPLOYEES HOLD YOU HOSTAGE
This candid book shares how to avoid being held hostage by employees.

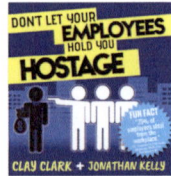

F6 JOURNAL
Meta Thrive Time Journal.

THE ENTREPRENEUR'S DRAGON ENERGY
The Mindset Kanye, Trump and You Need to Succeed.

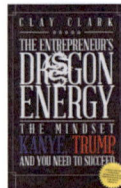

BOOM
The 13 Proven Steps to Business Success.

MAKE YOUR LIFE EPIC
Clay shares his journey and struggle from the dorm room to the board room during his raw and action-packed story of how he built DJConnection.com.

JACKASSARY
Jackassery will serve as a beacon of light for other entrepreneurs that are looking to avoid troublesome employees and difficult situations. This is real. This is raw. This is unfiltered entrepreneurship.

THE ART OF GETTING THINGS DONE
Clay Clark breaks down the proven, time-tested and time freedom creating super moves that you can use to create both the time freedom and financial freedom that most people only dream about.

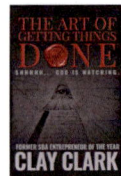

THRIVE
How to Take Control of Your Destiny and Move Beyond Surviving... Now!

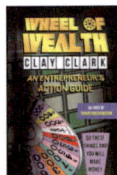

WILL NOT WORK FOR FOOD
9 Big Ideas for Effectively Managing Your Business in an Increasingly Dumb, Distracted & Dishonest America

WHEEL OF WEALTH
An Entrepreneur's Action Guide.

SEARCH ENGINE DOMINATION
Learn the Proven System We've Used to Earn Millions.

"SELF-MASTERY IS A DIFFICULT STEP, BUT A NECESSARY ONE. EOFIRE CAME TO LIFE BECAUSE I SET AND ACCOMPLISHED A GOAL, AND THAT WAS BUILT INTO A 7 FIGURE A YEAR BUSINESS BECAUSE I MASTERED PRODUCTIVITY, DISCIPLINE, AND FOCUS."

JOHN LEE DUMAS

(The founder and host of the *Entrepreneurs on Fire Podcast* who has been nice enough to be featured on the *Thrivetime Show* once and to interview me on his show twice.)

www.ingramcontent.com/pod-product-compliance
Lightning Source LLC
Chambersburg PA
CBRC090850210326
41597CB00007B/157